SHORT FIBER REINFORCED COMPOSITE MATERIALS

A symposium
sponsored by ASTM
Committee E-9 on Fatigue
in cooperation with
Society of Automotive Engineers and
American Society of Civil Engineers
Minneapolis, Minn., 14–15 April 1980

ASTM SPECIAL TECHNICAL PUBLICATION 772
B. A. Sanders, General Motors Corporation, editor

ASTM Publication Code Number (PCN)
04-772000-30

1916 Race Street, Philadelphia, Pa. 19103

NOTE

The Society is not responsible, as a body,
for the statements and opinions
advanced in this publication.

Printed in Baltimore, Md.
June 1982

Foreword

The Symposium on Short Fiber Reinforced Composite Materials, sponsored by ASTM Committee E-9 on Fatigue in cooperation with the Society of Automotive Engineers and the American Society of Civil Engineers, was held in Minneapolis, Minnesota, on April 14 and 15, 1980. B. A. Sanders, General Motors Corporation, has served as editor of this publication.

Related
ASTM Publications

Statistical Analysis of Fatigue Data, STP 744 (1981), 04-744000-30

Fatigue Crack Growth Measurement and Data Analysis, STP 738 (1981), 04-738000-30

Fatigue of Fibrous Composite Materials, STP 723 (1981), 04-723000-33

Joining of Composite Materials, STP 749 (1981), 04-749000-33

Test Methods and Design Allowables for Fibrous Composites, STP 734 (1981), 04-734000-33

Nondestructive Evaluation and Flaw Criticality for Composite Materials, STP 696 (1979), 04-696000-33

Design of Fatigue and Fracture Resistant Structures, STP 761 (1982), 04-761000-30

A Note of Appreciation
to Reviewers

This publication is made possible by the authors and, also, the unheralded efforts of the reviewers. This body of technical experts whose dedication, sacrifice of time and effort, and collective wisdom in reviewing the papers must be acknowledged. The quality level of ASTM publications is a direct function of their respected opinions. On behalf of ASTM we acknowledge with appreciation their contribution.

ASTM Committee on Publications

Editorial Staff

Contents

Introduction   1

Crack Propagation Modes in Injection Molded Fiber Reinforced
Thermoplastics—J. F. MANDELL, D. D. HUANG, AND
F. J. MCGARRY   3

Effect of Fiber Systems on Stiffness Properties of Chopped Fiber
Reinforced Sheet Molding Compound Composites—
D. C. CHANG   33

Effect of Processing Variables on the Mechanical and Thermal
Properties of Sheet Molding Compound—R. W. TUNG   50

Static and Fatigue Strength of Glass Chopped Strand
Mat/Polyester Resin Laminates—M. J. OWEN   64

An Investigation of Stress-Dependent, Temperature-Dependent,
and Time-Dependent Strains in Randomly Oriented Fiber
Reinforced Composites—E. M. CAULFIELD   85

Nondestructive Characterization of Chopped Glass Fiber Reinforced
Composite Materials—J. C. DUKE, JR.   97

Mechanical Behavior of Three Sheet Molding Compounds—
D. E. WALRATH, D. F. ADAMS, D. A. RIEGNER, AND
B. A. SANDERS   113

Vibration Characteristics of Automotive Composite Materials—
R. F. GIBSON, ANNA YAU, AND D. A. RIEGNER   133

Statistical Fracture Initiation in Randomly Oriented
Chopped-Mat Fiber Composites Subjected to Biaxial
Thermomechanical Loading— S. S. WANG AND T. P. YU   151

Elastic and Strength Properties of Continuous/Chopped
Glass Fiber Hybrid Sheet Molding Compounds—
N. S. SRIDHARAN   167

Dynamic Mechanical Characterization of Fiber Filled Unsaturated Polyester Composites—JON COLLISTER AND MICHAEL GRUSKIEWICZ   183

Fracture Characterization of a Random Fiber Composite Material—R. M. ALEXANDER, R. A. SCHAPERY, K. L. JERINA, AND B. A. SANDERS   208

Viscoelastic Characterization of a Random Fiber Composite Material Employing Micromechanics—K. L. JERINA, R. A. SCHAPERY, R. W. TUNG, AND B. A. SANDERS   225

Summary   251

Index   257

Introduction

This volume contains papers from the symposium on Short Fiber Reinforced Composite Materials held April 14 to 15, 1980, in Minneapolis, Minnesota. The symposium was sponsored by ASTM Committee E-9 on Fatigue in conjunction with the Society of Automotive Engineers and the American Society of Civil Engineers. It provided a forum on a rapidly growing area in composites technology: short fiber composites (SFC). These materials are growing in prominence in automotive, consumer appliance, and commercial business machine applications. They offer the advantage of weight reduction, design flexibility, energy savings, and high-volume processes for appearance and structural applications. In order to obtain the maximum advantages from SFC material systems, increasing amounts of research and development activities have been initiated at many companies, universities, and research institutes.

Emphasis in this symposium has been placed on SFC materials produced by SMC (sheet molding compound) compression molding and injection molding processes. A majority of the papers concentrate on automotive appearance and structural grade SMC material systems: SMC-R25, SMC-R50, and SMC-C/R. The potential of these materials in diverse automotive applications and the research being done generated considerable discussion among the conference attendees and presenters.

The papers in this volume, a selection of those presented at the symposium, cover a wide range of topics, including new material developments, test method development, mechanical and viscoelastic engineering properties, fracture behavior, and environmental effects. Thus this publication is one of the few compendiums of information available on SFC for high-volume process applications. These papers represent areas in which work is being done to provide the fundamental theoretical and experimental knowledge necessary to raise confidence in the use of SFC materials in more structurally demanding applications. In the past this type of work has been concentrated more on the relatively low-volume continuous fiber materials and processes used for aerospace applications.

Like their continuous fiber counterpart materials used in aerospace applications, SFC have many of the same needs with respect to design methodology development, mechanical behavior understanding, environmental effects, and test method development, among other developmental areas. The data presented at the symposium contribute to fulfilling these needs by adding to the base of knowledge needed to accelerate the use of SFC in automotive and other potential application areas.

This conference should be a forerunner of many on SFC material systems. Such forums will keep material engineers, design engineers, and researchers abreast of the issues being addressed and resolved in order to continue moving SFC technology further towards its full potential.

B. A. Sanders

General Motors Manufacturing Development, GM Technical Center, Warren, Michigan; editor

J. F. Mandell,[1] D. D. Huang,[1] and F. J. McGarry[1]

Crack Propagation Modes in Injection Molded Fiber Reinforced Thermoplastics

REFERENCE: Mandell, J. F., Huang, D. D., and McGarry, F. J.,**"Crack Propagation Modes in Injection Molded Fiber Reinforced Thermoplastics,"** *Short Fiber Reinforced Composite Materials, ASTM STP 772*, B. A. Sanders, Ed., American Society for Testing and Materials, 1982, pp. 3-32.

ABSTRACT: The modes of crack propagation are reported for injection molded short glass and carbon fiber reinforced thermoplastics. The matrices ranged from ductile to brittle, including Nylon 66, polycarbonate, polysulfone, poly(amide-imide), and polyphenylene sulfide; fiber contents were 30 or 40 percent by weight. The main crack is found to grow in a fiber avoidance mode, bypassing regions of agglomeration of locally aligned fibers. The local mode of crack tip advance varied with matrix ductility and bond strength. The fracture toughness and fatigue resistance of each material are related to the mode of crack growth.

KEY WORDS: short fiber composite materials, fracture, fatigue, crack propagation modes, injection molding

The work described in this paper is closely related to recent papers on the fracture toughness [1] and fatigue resistance [2] of glass and carbon fiber reinforced injection molded thermoplastics.[2] The materials used in these studies were reinforced with 30 to 40 percent by weight glass or carbon fibers; the matrices varied from ductile to very brittle. The fibers are oriented and broken down by the melt flow process, yielding a partially oriented system with variable fiber lengths. The distribution of fiber length, given in Ref 2 for each material, showed a predominance of fragments with ℓ/d_f (length-to-diameter) ratios of less than 20, and with the maximum fiber length less than 1 mm.

The fracture toughness in Ref *1* was measured by using notched tension

[1]Department of Materials Science and Engineering, Massachusetts Institute of Technology, Cambridge, Mass. 02139.
[2]The italic numbers in brackets refer to the list of references appended to this paper.

3

specimens of various sizes. For a given direction and location on the reinforced specimens, a single-value fracture toughness, K_Q, appeared to be
valid over most of the specimen size range [1]. Using the ultimate tensile
strength (UTS) from Type I bars as specified in ASTM Test for Tensile Properties of Plastics (D 638), the crack-tip zone over which the UTS was exceeded was calculated as

$$r_c \cong \frac{1}{2\pi} \left(\frac{K_Q}{\text{UTS}} \right)^2 \qquad (1)$$

Figure 1 shows that the calculated critical zone size correlates in
magnitude and trend with the length of the longer fibers for each material,
taken for convenience as the length exceeded by 5 percent of the fibers. The
fracture specimens in this case had a crack length of 2.03 cm. The data suggest that for a given UTS, the fracture toughness is determined by the fiber
length, regardless of fiber material, matrix ductility, or fiber/matrix bond
strength. Table 1 lists the values of K_Q and UTS for each material; it should
be noted that the K_Q-values for the three tougher matrices—N66, PC, and
PSUL—do not appear to satisfy the plastic zone size requirement of ASTM

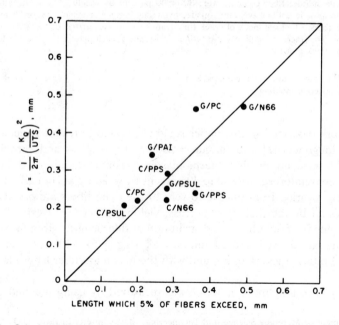

FIG. 1—*Calculated critical zone radius versus fiber length exceeded by 5 percent of fibers: various carbon and glass fiber reinforced thermoplastics with 30 to 40 percent fiber by weight* [1].

TABLE 1—*Strength, fracture toughness, and fatigue properties.*[a]

Property	Reinforcement	Matrix				
		N66	PC	PSUL	PPS	PAI
K_Q, MN m$^{-3/2}$	none[b]	10.9	8.4	4.7	0.8	...
	glass	9.9	8.7	6.5	7.0	9.4
	carbon	9.4	7.5	7.2	6.6	...
UTS, MPa	none[c]	74	72	77	35	140
	glass	181	161	158[d]	181	203
	carbon	256	203	197	156	231
Percent loss of UTS/decade fatigue cycles	glass	10.9	11.4	...	10.9	10.5
	carbon	9.9	11.5	11.1	7.8	9.5

[a] After Refs 1 and 2.
[b] K_Q data for N66 and PC do not satisfy validity criteria of ASTM E 399.
[c] Values for N66, PC, and PSUL are yield stress.
[d] Estimated value [1].

Test for Plane-Strain Fracture Toughness of Metallic Materials (E 399). These matrices are, however, clearly very tough compared with PPS.

The *S-N* (maximum stress versus log cycles to failure) fatigue behavior of these same materials was reported in Ref 2. Except for the N66 matrix systems, all *S-N* curves appeared linear over the entire loading range in tension-tension ($R = 0.1$) fatigue, where R = maximum stress/minimum stress. The glass-reinforced materials all lost between 10.5 and 11.4 percent of their strength per decade of cycles (Table 1), close to the value observed for other glass-reinforced materials such as sheet molding compounds (SMC) and unidirectional-ply laminates [3]. The carbon-reinforced materials showed more variation in fatigue resistance, best for the brittle matrices, and worst for the ductile ones.

In each of the above studies, the mode of crack growth was observed to involve single macroscopic cracks that characteristically grow around the fibers on a local scale. The work described in this paper was intended to define the crack growth mechanisms for each material in an attempt to explain the fracture and fatigue behavior.

Experimental Methods

The matrix materials were all engineering thermoplastics, including semicrystalline Nylon 66 (N66), amorphous polycarbonate (PC), and polysulfone (PSUL), as well as two higher temperature plastics, amorphous poly(amide-imide) (PAI) and semicrystalline polyphenylene sulfide (PPS). The unreinforced PPS is very brittle at room temperature, while the N66, PC, and PSUL all neck and draw in uniaxial tension. The PAI shows slight ductility in tension, but is most notable for a high elastic strain capability and high tensile strength. The fibers used were E-glass or PAN-based carbon with a modulus of approximately 207 GPa. The fiber surface treatments or coupling agents where present are proprietary formulations of the material suppliers [1,2]. The fiber content was 40 percent by weight, except for the PAI systems, which were 30 percent; these are high values for this class of materials.

The specimens used in this study were end-gated ASTM D 638 Type I tensile bars [1,2]. The fiber orientation was variable, but the dominant direction was along the specimen length. Cracks were grown across the width, normal to the dominant fiber orientation. The loading was either tension-tension fatigue as described in Ref 2, or wedge loading at a precut notch in a microscope loading fixture. Earlier fatigue work [2] was done with the same lot of specimens, while the fracture work [1] used larger plaques, but with the crack in the same direction relative to the fiber orientation. All tests were conducted in an air-conditioned laboratory atmosphere.

Results and Discussion

Fiber Avoidance Mode of Crack Growth

Cracks in all the reinforced materials propagated in such a manner as to avoid most of the fibers, as suggested in previous studies [1–3]. Figure 2 illustrates this mode in low magnification micrographs which, at this scale, are typical of all materials studied. The planar view gives the most detailed information about the crack path. The specimen in this case has been metallographically polished on the surface with sufficient care to avoid any damage such as fiber cracking. After the surface is prepared and inspected at high magnification, a crack is grown in the desired direction by forcing a wedge into a precut edge notch. The crack growth is generally stable, so that the crack tip can be observed at the desired magnification in a light microscope as it propagates.

Figure 2 and subsequent micrographs indicate that the fibers tend to agglomerate into small groups, roughly parallel in orientation. The length of these local agglomerations tends to be similar to the length of the longest fibers in the region. However, it should be noted that a planar surface seldom exposes the fiber over its entire length, and the fibers appear somewhat shorter in these micrographs than they actually are. They usually either are cut by the surface or plunge back into the interior after some exposed distance. Observation of all reinforced materials at low magnification (Fig. 2) reveals that the crack follows a path that avoids these agglomerated fiber groups as much as possible. This leaves a zig-zag appearance, as shown. Interactions of the crack with more isolated fibers not associated with the agglomerations results in some debonded and pulled out or broken fibers on the fracture surface. Studies of the fracture surface, commonly conducted in

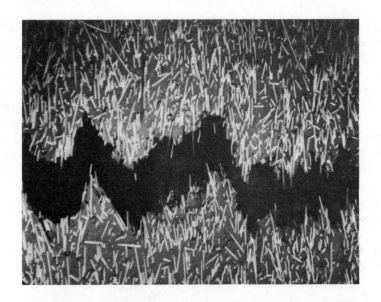

FIG. 2—*Low-magnification views of fiber-avoidance mode of crack growth in carbon fiber reinforced polycarbonate (top, planar view in light microscope) and Nylon 66 (bottom, SEM fractograph);* $d_f \cong 10 \ \mu m$.

the scanning electron microscope (SEM), fail to reveal information about the interactions of the crack with the fiber agglomerations that appear to dictate its path for this group of materials.

The SEM view at low magnification in Fig. 2 does indicate that the planar zig-zag path on the surface is repeated in the thickness direction as well. Thus a crack front will not define a flat plane, but will be moving locally in different directions at different positions through the thickness, depending on the local fiber arrangement. The various parts of the crack front must finally join together to form a continuous zig-zag path by breaking away any material between the local growth regions. On a given plane, such as that in Fig. 2 (top), the local crack must frequently be redirected along the main crack path rather than continuing along a direction more nearly parallel to the fibers to give a single macroscopic crack identity. On a given plane, the crack path is strongly influenced by the development of the crack on adjacent planes above and below that level. The implications of the fiber avoidance mode for the crack resistance of the material will be discussed later.

Local Mechanisms of Crack Advance

Fracture toughness may be predictable simply from the gross nature of the fiber avoidance mode, as will be discussed later. However, fatigue resistance appears to be sensitive to the local mechanisms of crack advance. This section will present observations of the mechanisms involved for each material as viewed during static loading in the microscope and after fatigue loading in a servohydraulic machine. It is difficult to obtain clear micrographs of what is readily observed in practice, and then only a sampling of hundreds of micrographs can be presented here due to length limitations. While most of the micrographs presented are planar surface views of static cracks grown in a stepwise fashion by loading increments, an attempt is made to relate these results to observations of the crack on the interior of the specimen and to fatigue cracks.

Figures 3 and 4 show the local mode of crack growth for the very brittle PPS matrix with glass and carbon fibers. The details of each are surprisingly different. The development of the crack tip zone is shown for the glass system in Fig. 3 at four successively higher loads. The crack path can be correlated from one micrograph to the next by the arrow locations. There is a region ahead of the main crack in which small separate cracks are formed, mostly around fiber ends. These cracks coalesce at higher loads to form the main crack. This mode results in cracks primarily in the matrix and across fiber ends, with little fiber pullout. Occasional fibers are broken by the crack, such as the one indicated by the lower arrow in Fig. 3d. The PPS system tends to contain many voids, which appear as dark circles in Fig. 3.

The same matrix with carbon fibers in Fig. 4 shows much more tendency to debond and pull out fibers. Little or no local cracking is observed around the main crack tip (Fig. 4d). The crack in the matrix appears to grow past debonded fibers, and then extract them from the matrix as it opens. However, the crack still grows around the major agglomerations of fibers as described earlier. Many of the debonded fibers appear to slide uniformly out of the matrix, as shown by the lower arrow in Fig. 4a. Such sliding requires that the main crack open along the fiber axis, since the fiber is rigid and does not distort significantly. The upper arrow in Fig. 4a indicates a fiber at another angle. This fiber is oriented in a different direction, and cannot slide out of its matrix sheath with the main crack opening in the direction dictated by the fibers on either side and possibly other fibers in the interior. Being on the surface, this fiber could pull free to accommodate the situation, but some such fibers are observed to break if they are bent across a crack that opens in a direction other than along the individual fiber axis. Such local variations in fiber orientation may contribute significantly to crack resistance, requiring fiber failure or matrix fragmentation before the main crack can open, even with the relatively small deviation in fiber alignment shown in Fig. 4a. It appears that a major difference between the glass and carbon systems may be a much lower bond strength in the carbon system, but direct bond strength measurements have not been made.

Figure 5 shows a distinctly different local mode of extension for the carbon-reinforced polycarbonate, which is typical of most of the ductile matrix systems. The main crack tip shown in Fig. 5d is actually a local continuous band of yielding and necking, with some fiber pullout. Further back along the main crack (Fig. 5c) the yielded material tears apart in some regions, and finally the whole yielded path tears apart to form a real crack. The arrows (lower right) in Figs. 5a and 5b show the same spot before and after crack growth. The region ahead of the crack contains many locally yielded areas, especially near fiber ends, such as at the arrow. These coalesce to form the main continuous yielded zone, which later tears apart to form the crack. As in the PPS/glass, this coalescence mode allows the crack to seek out the path of least resistance, with occasional fibers crossed by the main yield zone extracted from the matrix. Figure 6 shows the same features for carbon-reinforced polysulfone and Nylon 66.

The ductile matrices with glass fibers show a similar mode of local ductile crack growth, as in Fig. 7b for the Nylon 66 matrix. A combined mode is evident for the glass reinforced polycarbonate in Fig. 7a. Widespread local cracks or crazes are formed ahead of the main crack, mostly near fiber ends. However, close to the main crack tip local necking and yielding appear to dominate. In each of the glass reinforced systems, fibers were occasionally observed to break where they bridged the main crack or yield band; fiber failure was rarely observed with carbon fibers.

The poly(amide-imide) matrix does not neck and draw in uniaxial tension, but shows considerable local ductility in the composite. Figure 8 shows a crack tip and subsequent growth for the carbon reinforced system. The car-

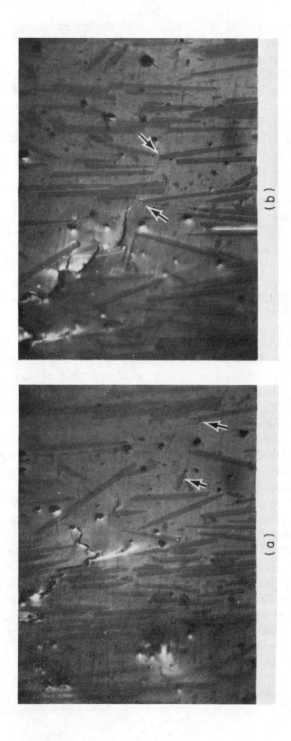

FIG. 3—Crack tip in glass-reinforced polyphenylene sulfide under increasing load (a) to (d); arrows indicate path crack will follow: $d_f \cong 11 \ \mu m$.

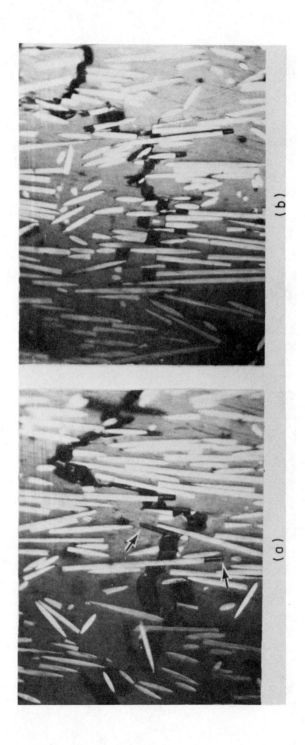

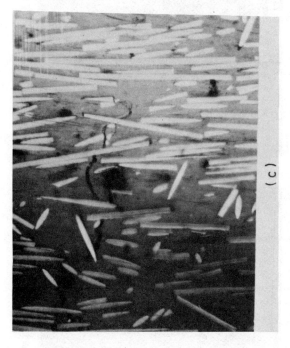

(c)

FIG. 4—*Various positions along crack in carbon-reinforced polyphenylene sulfide:* $d_f \cong 10 \ \mu m$.

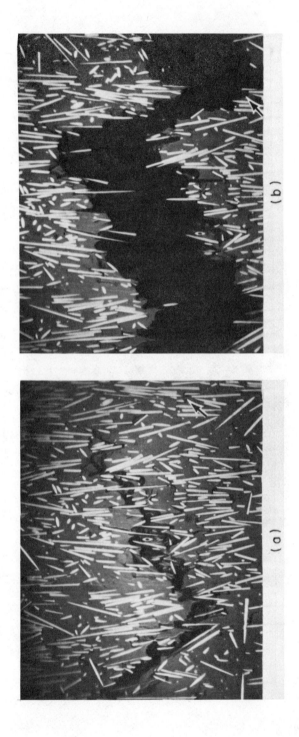

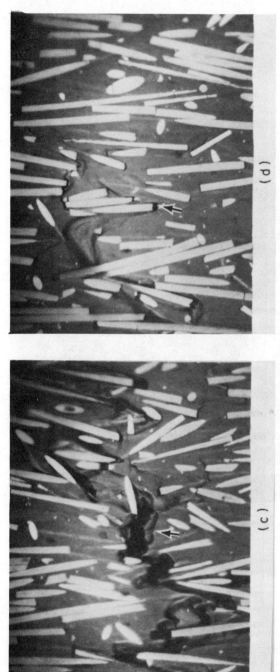

FIG. 5—*Crack in PC/C: (c) and (d) are magnified view of crack in position (a); (b) is same area as (a), but after crack growth: crack path indicated by arrows in (a) and (b); $d_f \cong 10 \mu m$.*

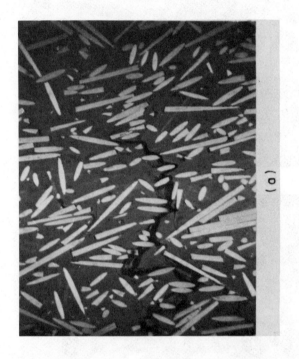

(a)

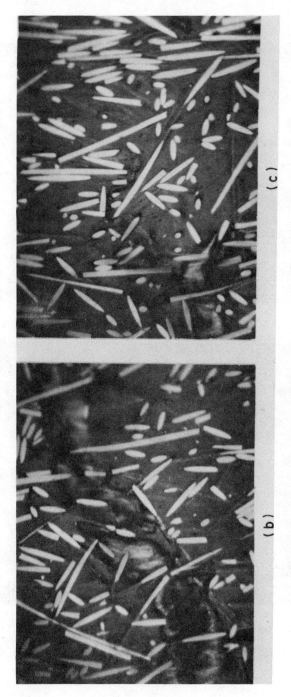

FIG. 6—*Cracks in carbon-reinforced polysulfone* (a) *and in Nylon 66 along crack* (b) *and at tip* (c); $d_f \cong 10 \ \mu m$.

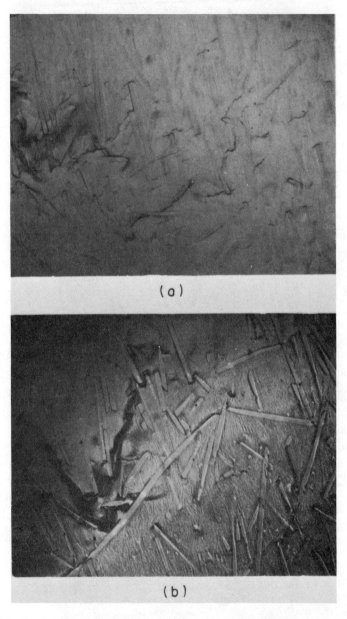

FIG. 7—*Crack tip in glass-reinforced polycarbonate* (a) *and Nylon 66* (b); $d_f \cong 11 \ \mu m$.

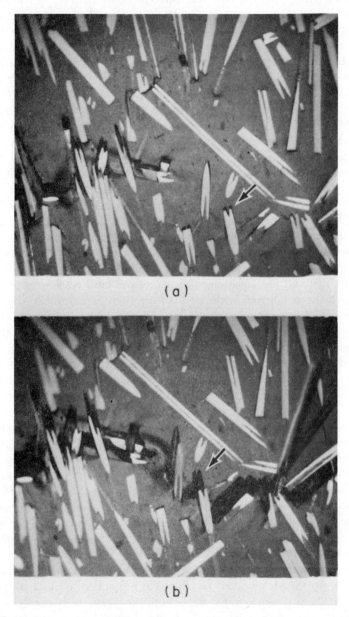

FIG. 8—*Crack tip* (a) *in carbon-reinforced poly(amide-imide) and same region after crack extension* (b); *arrows indicate crack path;* $d_f \cong 15 \mu m$.

bon fibers in this material have a kidney-shaped cross-section, giving a split appearance when polished part way through. Local tearing of the yielded material is evident to the left of the arrow in Fig. 8*b*. Figure 9 shows the development of damage at a notch root in the glass-reinforced system. The arrows in Figs. 9*a* and 9*b* trace the development of fiber debonding (which occurred first), yielding, and fiber fracture. (The fiber at the immediate notch boundary was cracked during notching.) The crack path (Fig. 9*c*) shows a broken fiber (arrows), and debonded, pulled-out fibers. Yielding was very localized in the PAI systems, but the brittle cracking observed in glass/PPS around fiber ends was not evident. Instead, the high strain capacity of the matrix appeared to result in fiber debonding and fracture, along with some local yielding.

Differences in the local mode of crack growth are also observed in SEM micrographs of the fracture surfaces. The carbon-reinforced systems in Fig. 10 show variations from brittle matrix fracture for PPS to very ductile for N66, with some ductility evident for PC and PAI. The same sequence with glass fibers in Fig. 11 (lower magnification than Fig. 10) shows similar effects. However, relatively few fibers are observed on the PPS surface, reflecting the local mode of formation of the crack in Fig. 3.

Figures 12 to 14 are an attempt to define the local crack growth modes observed using schematics with an arbitrary fiber arrangement. Figure 12 represents the local cracking, crack coalescence mode of PPS/glass, with occasional fiber failure. Figure 13 represents the PPS/carbon system with poor bonding, showing frequent fiber debonding and pullout. Figure 14 represents the ductile matrix systems, with local yielding ahead of the crack followed by coalescence into a main yielded zone which eventually tears to form a crack. Here, the dark areas at the fiber ends represent yielding, while the light gaps represent separation.

It is important to note that these results are primarily from examination of cracks on the surface. Polishing of the material after crack growth to observe the interior tends to smear the matrix detail. However, such observation, coupled with the observation of cross sections of crack tips, indicates that the surface results generally also apply to the interior. The brittle matrix growth mechanisms are unchanged. The ductile systems cannot form a continuous local necked zone, but form voids between drawn regions. The ductility in the interior is also evident in the SEM micrographs. Much of the ductility appears to be in response to shear stresses resulting from relative movement of adjacent groups of fibers, rather than plane stress effects at the surface. Investigation by SEM revealed no significant increase in drawn, yielded material at the surface, as might be expected of plane stress effects.

The effect of fatigue cycling on the mode of crack growth was determined by growing fatigue cracks across previously polished specimens of each material, then observing under slight load in the microscope. Although damage such as local cracking near fiber ends may have been slightly more widespread in fatigue, no significant differences from static cracks in the mode of growth or the fracture surface could be identified. Some isolated regions of fatigue striations on cracks around fiber ends were reported in Ref 2, but it was generally not possible to distinguish a fatigue crack from a static crack.

Interpretation of Fracture and Fatigue Data

Fracture Toughness—The micrographs clearly give several possible alternatives for explaining the fracture toughness. The traditional approach would concentrate on the fiber pullout friction and debonding energy [4], as well as ductile flow where present. However, the empirical findings represented in Fig. 1 suggest a criterion that does not depend directly on the local mechanisms of crack resistance. All data in Fig. 1 were for cracks greater than 2 cm in length, at least 40 times the long fiber length or calculated crack-tip zone radius. Thus all the long fibers and agglomerations should be embedded in the singular crack tip stress field, and the calculation of K_I at fracture should be valid [5].

The fracture criterion that has been suggested for the fiber avoidance mode [1] postulates that the crack should be able to propagate if the local stress reaches the UTS at a critical distance, r_c, from the crack tip, similar to the length of the longer fibers in the distribution, termed ℓ_f^* (taken in Fig. 2 for convenience as the length exceeded by 5 percent of the fibers). For the fiber arrangement assumed in the schematics, this distance would be represented as in Fig. 15. Since this condition is reached all through the thickness, the zig-zag crack front evidenced in Fig. 2 should be able to grow around any long fiber or agglomeration and move in a macroscopic sense.

Calculation of the critical crack tip dimension mixes microstructure with continuum mechanics, and is clearly very approximate in nature. Ignoring directional effects, the local stresses around the crack tip are approximately [5]:

$$\text{local stresses} \cong \frac{K_I}{(2\pi r)^{1/2}} \tag{2}$$

where r is the distance from the crack tip. The crack will propagate by this criterion when the local stress reaches the UTS at a critical radius r_c equal to the length of the longer fibers or agglomerations:

$$r_c = \ell_f^* \cong \frac{1}{2\pi} \left(\frac{K_I}{\text{UTS}} \right)^2 \tag{3}$$

and at this condition K_I will be the fracture toughness K_Q, so that

$$K_Q \cong \text{UTS}(2\pi\ell_f^*)^{1/2} \tag{4}$$

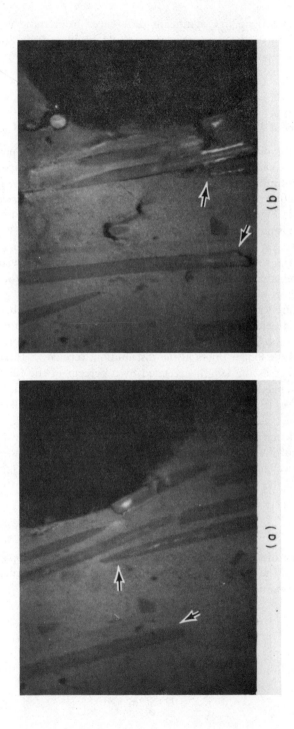

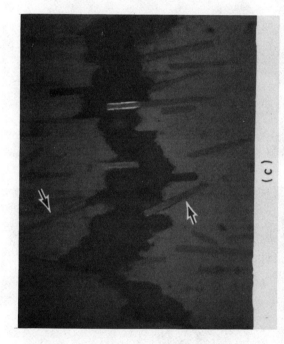

FIG. 9—Glass-reinforced poly(amide-imide) showing damage development at notch root at low (a) and high (b) loads, and crack path (c); $d_f \cong 11$ μm.

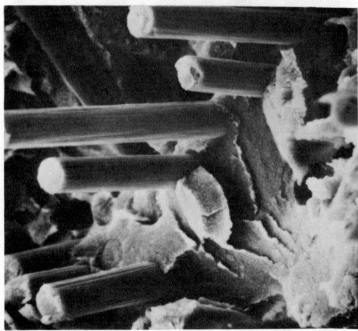

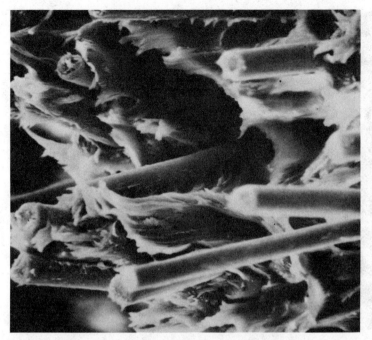

FIG. 10—*Fracture surfaces of carbon-reinforced matrices (clockwise from upper left): polyphenylene sulfide, poly(amide-imide), Nylon 66, polycarbonate;* $d_f \cong 10 \ \mu m$.

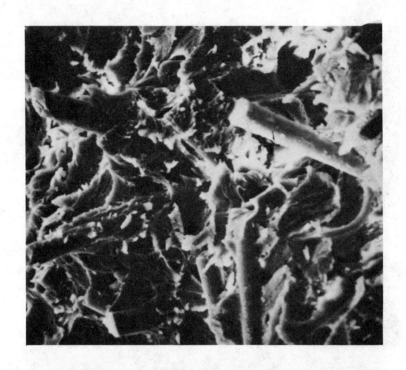

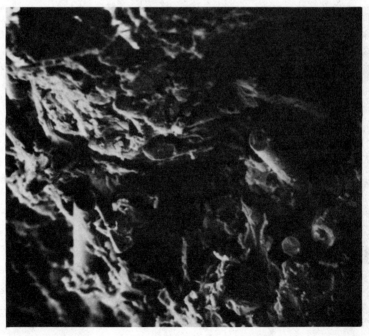

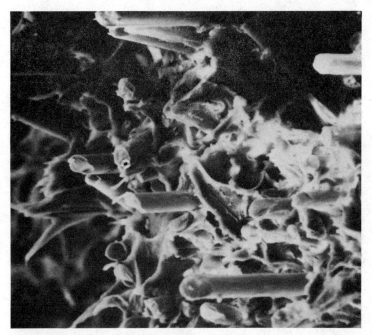

FIG. 11—*Fracture surfaces of glass-reinforced matrices (clockwise from upper left): polyphenylene sulfide, poly(amide-imide), Nylon 66, poly-carbonate;* $d_f \cong 11 \ \mu m$.

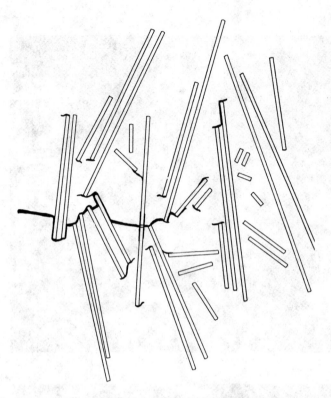

FIG. 12—*Schematic of local mode of crack growth in well-bonded, brittle matrix, glass fiber materials.*

This relationship appears to be in agreement with the data in Fig. 1. The great variation in matrix toughness and apparent bond strength evident for this group of materials seems not to greatly influence the fracture toughness. What influence the constituent and interface properties do have enters through the UTS, and even that is modest (Table 1). The UTS is also likely to depend on fiber length for these materials [4].

Fatigue Effects—Fatigue appears to involve the gradual development and extension of cracks in all these materials [2]. While the fracture toughness does not appear to be influenced by the local details of crack advance, cycling the material at less than the critical load may give more sensitivity. The similarity of the rate of degradation in fatigue of the glass reinforced systems to that observed for the chopped strand and continuous fiber glass/epoxy composites discussed earlier [2,3] indicates that fatigue may be a fiber-sensitive property for glass-reinforced materials. The observation of occasional glass fiber failure in each material supports this view. By implication,

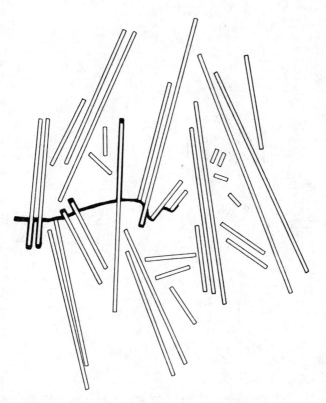

FIG. 13—*Schematic of local mode of crack growth in poorly bonded, brittle matrix materials.*

the few fibers that are not avoided or pulled out by the crack, but finally must be broken, may determine the fatigue resistance.

Carbon fiber composites generally show very little degradation in fatigue for long fiber unidirectional materials [6]. Carbon fiber failures were very rarely observed in this study in fatigue or static crack growth. These observations appear to leave only the matrix and interface to influence the fatigue of the carbon reinforced materials, and a significant sensitivity to the matrix material was observed in Ref 2 (Table 1). The best fatigue resistance is provided by the brittle PPS matrix system, which tends to have many debonded fibers bridging the crack. Their effect is to provide force transfer across the crack faces by friction or by the interlocking mechanism of fibers with different orientation described earlier. A tension test of a specimen cut from a fatigue crack tip was reported in Ref 2 to support a stress of 24 MPa. Fatigue of the ductile matrix materials appears to involve cyclic failure of the yielded matrix ahead of the crack tip, and poorer fatigue resistance is obtained.

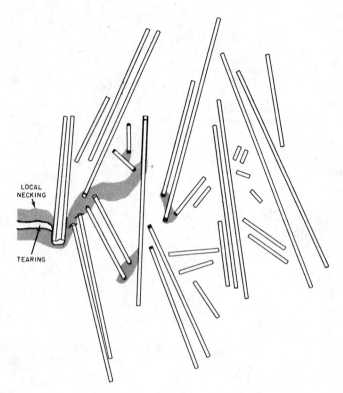

FIG. 14—*Schematic of local mode of crack growth in ductile matrix material.*

Discussion

The conventional view of the strength and crack resistance of short fiber composites has been closely tied to the concept of the critical fiber length ℓ_c [4]. The critical length is that which is sufficient to allow the stress transferred through shear at the interface to load the fiber to failure. This model involves an isolated fiber oriented parallel to the applied uniaxial tensile stress. The impact and crack resistance is often viewed as deriving from the work of extracting fibers or pieces of fibers shorter than ℓ_c from the crack surfaces [4,7]. It is difficult to envision the usefulness of the ℓ_c model for the relatively high fiber content, poorly aligned fiber cases studied here, owing to the following observations:

1. Of the fibers that break, most appear to be bent across the crack, which does not open along their axes.
2. Cases in which the crack avoids nearly all the fibers, such as PPS/glass, have strength and fracture toughness comparable to the other materials.
3. Since the fibers are not aligned in parallel, they cannot all slide

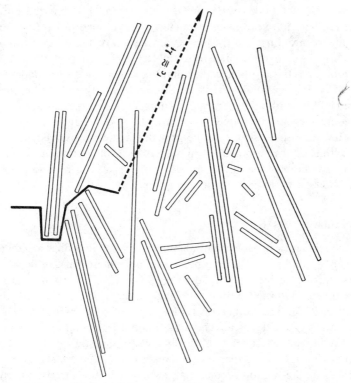

FIG. 15—*Model for critical crack-tip zone radius for fiber-avoidance mode of fracture.*

smoothly out of the matrix as the crack opens in one general direction. The fibers appear too stiff to bend appreciably, as may occur in metal wire reinforced concrete [8].

4. The stresses applied to the fiber are complicated greatly by the cracking at the fiber ends, partial debonding, contact with adjacent fibers, etc.

The results suggest several points of interest in materials development. Firstly, longer fibers are better for toughness as well as strength as long as the same modes of crack extension occur. It is doubtful that this observation would hold for significantly longer fibers, since they would be more likely to break. For fatigue resistance, little change appears likely for glass-reinforced materials, which lose approximately 10 percent of their UTS per decade of cycles under these loading conditions. For graphite-reinforced materials, a brittle matrix, poorly bonded system seems to give the best fatigue performance, but with some loss in UTS, compared with a ductile or well-bonded material.

Throughout this work the crack was grown normal to the dominant fiber orientation. Crack growth more nearly parallel to the fibers gives much lower values of K_Q for both brittle and ductile systems [1], but the observations presented here have not been examined in detail for this case. The relatively high fiber content used in this study may also have had a significant effect on the results. Micrographs like those in Ref 7 suggest that the matrix crack may grow on a flatter plane rather than in a fiber-avoidance mode if the fiber content is significantly lower; if so, this could change all of the observations made in this study. Higher rate loading or lower temperatures will likely cause changes in the ductile systems. A mode change to brittle matrix behavior after some slow, ductile growth has been observed in very recent tests with carbon-reinforced Nylon 66 at a lower fiber content under the same loading conditions used in this study. At high rates this mode change has also been observed at 40 percent fiber content.

Conclusions

The main crack in each material appeared to follow a fiber avoidance mode, growing around the longer fibers and agglomerations of locally aligned fibers. A fracture criterion based on reaching the UTS at a distance from the crack tip equal to the length of the longer fibers is in agreement with fracture toughness data. The fatigue resistance appears to be more influenced by the local details of crack tip advance. Occasional fiber failure appears to dominate the fatigue resistance of glass fiber systems, resulting in behavior similar to other glass reinforced plastics. The matrix and interface are more important for the graphite-reinforced materials, where a brittle matrix, poorly bonded system gave the best performance and a distinctive mode of crack advance.

Acknowledgments

This work was sponsored by the Lord Corporation, and materials were molded and supplied by its Keyon Materials facility. The work of George Normann in specimen preparation is also appreciated.

References

[1] Mandell, J. F., Darwish, A. Y., and McGarry, F. J. in *Test Methods and Design Allowables for Fibrous Composites, ASTM STP 734*, C. C. Chamis, Ed., American Society for Testing and Materials, 1981, pp. 73-90.
[2] Mandell, J. F., Huang, D. D., and McGarry, F. J., *Polymer Composites*, Vol. 2, 1981, p. 137.
[3] Mandell, J. F., *Polymer Composites*, Vol. 2, 1981, p. 22.
[4] Kelly, A., *Strong Solids*, 2nd ed., Clarendon Press, Oxford, 1973, p. 157.
[5] Knott, J. F., *Fundamentals of Fracture Mechanics*, Butterworths, Boston, 1973, p. 133.
[6] Dharan, C. K. H., "Fatigue Failure Mechanisms in Pultruded Graphite-Polyester Composites," in *Failure Modes in Composites II*, J. N. Fleck and R. L. Mehan, Eds., The Metallurgical Society of AIME, New York, 1974, p. 144.
[7] Ramsteiner, F. and Theysohn, R., *Composites*, Vol. 10, 1979, p. 111.
[8] Naaman, A. E., Argon, A. S., and Moavenzadeh, F., *Cement and Concrete Research*, Vol. 3, 1973, p. 397.

D. C. Chang[1]

Effect of Fiber Systems on Stiffness Properties of Chopped Fiber Reinforced Sheet Molding Compound Composites

REFERENCE: Chang, D. C., "**Effect of Fiber Systems on Stiffness Properties of Chopped Fiber Reinforced Sheet Molding Compound Composites,**" *Short Fiber Reinforced Composite Materials, ASTM STP 772,* B. A. Sanders, Ed., American Society for Testing and Materials, 1982, pp. 33–49.

ABSTRACT: The effect of such fiber characteristics as weight content, modulus, and density on the stiffness properties of a chopped fiber reinforced sheet molding compound (SMC) composite are analytically evaluated. Comparisons are made among SMC composites reinforced by four commercially available fiber systems, glass, Kevlar, carbon, and high-modulus carbon.

KEY WORDS: sheet molding compounds, chopped fibers, composite materials, stiffness properties, automotive applications

Automotive structures are distinguished by the need to mass-produce reliable and geometrically complex parts with minimum labor intensity and low process time. The chopped-fiber reinforced sheet molding compound (SMC) is particularly suited for such manufacturing requirements and is likely to find greater application than directional types of composites in replacement of steel for car structures. One significant shortcoming of the glass-fiber reinforced SMC composite that is currently used in car production is its low stiffness relative to steel. The low-modulus glass-SMC results in modest mass savings over steel [1],[2] but may require the part to be made of relatively thick gage. Consequently, the part will need longer cure time and this will result in higher manufacturing cost.

---

[1] Senior Staff Research Engineer, Engineering Mechanics Dept., General Motors Research Laboratories, Warren, Mich. 48090.
[2] The italic numbers in brackets refer to the list of references appended to this paper.

For the SMC-composite to be more competitive and have broader application than other alternative materials for replacing steel, it is essential that the
stiffness properties of SMC be improved over current glass-SMC composites.
In this study, we will examine to what extent the fiber phase affects the stiffness of the SMC composite. We will analytically evaluate the degree of influence resulting from changing fiber content, fiber modulus, and fiber density
on the composite stiffness. Finally, SMC composites reinforced by four commercially available fibers such as glass, Kevlar, carbon, and high-modulus
carbon will be compared. The analytical method presented by Chang and
Weng [2] for chopped-fiber reinforced composites is used to compute the effective stiffness properties of the SMC composites.

Material Compositions

There are two kinds of chopped-fiber reinforced SMC: unfilled and filled.
The basic constituents of an SMC composite are resin, filler, fiber, and small
amounts of additives. The commonly used fibers of automotive SMC are
coated S-glass filaments combined as rovings and chopped to 25 mm in
length. The chopped fibers are randomly dispersed in a filled or unfilled
thermosetting resin paste (polyester) to improve strength, stiffness, and dimensional stability of the composite. For a filled SMC, a large fraction of
low-cost filler is used to reduce the overall cost of the composite. The most
commonly used filler for automotive-type SMC is finely ground calcium carbonate ($CaCO_3$). Other miscellaneous additives comprise only a very small
weight fraction of the composite. Thus, in computing the effective stiffness
properties of a typical SMC composite, the direct contribution from these additives is ignored and their weights are assigned to the resin weight. By doing
so, an unfilled SMC can be simply identified as a two-phase composite
(fiber/resin), and a filled SMC can be identified as a three-phase composite
(fiber/filler/resin).

In this study, other commercially available fibers such as carbon, high
modulus carbon, and Kevlar will be evaluated as potential candidate reinforcements for SMC composites. To establish a common base for comparison, we generate all SMC composites, filled or unfilled, based on a typical
commonly used chopped-glass SMC composition [3]. All SMC composites
considered have the same resin (polyester) and filler (calcium carbonate) systems. The typical constituent properties used in this study are listed in Table 1.

Automotive-type SMC is generally designated by its fiber weight fractions;
for example, SMC-50 contains 50 percent chopped fiber by weight. The
amount of resin needed to ensure adequate wetting of fibers and filler particles is, however, established on a minimum resin volume basis. For an unfilled SMC, since the change in fiber weight fraction is balanced by the
change in resin weight fraction, the maximum permissible fiber weight fraction is then governed by the minimum resin volume needed and is different

TABLE 1—*Constituent properties of SMC composites.*

Composite Constituents	Young's Modulus (E), GPa	Poisson's Ratio (ν)	Density (ρ), g/cm^3
Polyester (resin)	3.5	0.45	1.12
Calcium carbonate (filler)	41.4	0.21	2.77
Glass fiber	72.4	0.22	2.54
Kevlar fiber	131.0	0.25	1.44
Carbon fiber	227.0	0.25	1.72
High-modulus carbon fiber	482.0	0.25	1.90

for different types of fibers. In the case of automotive-type filled glass-SMC, the change of glass fiber weight is offset by an equal change of filler weight with the resin weight being kept constant ($\sim$34 percent) [2]. Because the densities of glass fiber and filler (calcium carbonate) are fairly close (2.70 versus 2.54), the corresponding resin volume of all filled glass SMC also remains almost constant ($\sim$55 percent). This ensures adequate wetting for each composition, and no more resin is used than is necessary, because it costs more than the filler.

When fibers that are much lighter than the filler are used to reinforce a filled SMC composite, the change of fiber content must accompany changes in both resin and filler weight fractions in order to maintain adequate wetting. This is so because for a given fiber weight fraction, lighter fibers result in higher fiber volume fraction. Therefore, in order to ensure adequate wetting for all filled SMC (same as for glass SMC) with different types of fibers, we must compose the filled SMC on an equal resin volume basis and determine the balanced filler and resin weight fractions for each given fiber weight fraction. Detailed compositions will be discussed later in the paper.

Effective Stiffness Properties

The analytical method developed by Chang and Weng [2] is used to determine the stiffness properties of both filled and unfilled SMC composites. The analytical results of Ref 2 compare well with experimental results for both filled (glass/calcium carbonate/polyester) and unfilled (glass/polyester) SMC composites. The basic approach of the method is that the resin and filler are combined to form an "effective" matrix and the resulting matrix is then combined with the chopped fibers to form the final composite. In this approach the assumptions imposed on modeling the two-phase and three-phase composites are that:

1. Each phase of the constituents is homogeneous and isotropic.
2. Both fibers and filler particles are uniformly dispersed, and the fibers are randomly oriented in the resin.

3. The fibers lie in the plane of the sheet.
4. Void effects are neglected.
5. The interfacial bond between the constituents is perfect.

Thus the resin/filler two-phase composite can be regarded as quasi-homogeneous and quasi-isotropic, and the resin/filler/fiber three-phase composite as quasi-homogeneous and, in the plane of the sheet, quasiisotropic. In other words, we have assumed the existence of "effective" or macroscopically averaged mechanical properties for both phases of the composite. This, of course, is a reasonable approximation only when dealing with the average behavior of a body containing many fibers or inclusions that are small in comparison with the gross dimensions of the body or region of interest, and when the size of filler particles is small in comparison with the smallest dimension of the fibers.

The idealized geometric model upon which the theoretical results are based is that of the sphere model [4] for particulate fillers and the cylinder model [5] for chopped fibers. Further, it is assumed that the fiber aspect ratio is sufficiently large so that the end effects of the chopped fibers can be ignored. For completeness, we have summarized the stiffness equations for both the resin/filler mixture and the fiber/matrix mixture as follows [2].

When the resin and filler are mixed to form the matrix phase for the final composite, the effective bulk modulus (K_m) and shear modulus (G_m) of the (filled) matrix are

$$K_m = K_r + (K_p - K_r) \frac{(4G_r + 3K_r) v_p^*}{4G_r + 3K_p + 3(K_r - K_p) v_p^*} \tag{1a}$$

$$G_m = G_r \left\{ 1 + \frac{15(1 - v_r)(G_p - G_r) G_r v_p^*}{(7 - 5v_r) G_r + 2(4 - 5v_r)[G_p - (G_p - G_r) v_p^*]} \right\} \tag{1b}$$

where subscripts m, r, and p refer to matrix, resin, and particulate filler, respectively, v_r is the Poisson's ratio of the resin, and

$$v_p^* = \frac{v_p}{v_p + v_r} \tag{2}$$

with v_p^* being the effective filler volume content of the matrix phase and v_p and v_r being the corresponding volume fractions of the final composite. The elastic modulus and Poisson's ratio, which are the effective stiffness properties needed to represent the filled matrix, can thus be derived as

$$E_m = \frac{9K_m G_m}{3K_m + G_m} \tag{3a}$$

$$v_m = \frac{3K_m - 2G_m}{2(3K_m + G_m)} \tag{3b}$$

Once the effective stiffness properties of the filled-matrix phase are known, a three-phase composite such as a fiber-reinforced filled SMC can be represented as a fiber/matrix mixture. The effective elastic moduli (E_c and v_c) of such a mixture can be expressed as

$$E_c = \frac{1}{u_1}(u_1{}^2 - u_2{}^2) \tag{4a}$$

$$v_c = \frac{u_2}{u_1} \tag{4b}$$

where the subscripts c refers to the final composite, and

$$u_1 = \frac{3}{8} E_{11} + \frac{G_{12}}{2} + \frac{(3 + 2v_{12} + 3v_{12}{}^2)G_{23}K_{23}}{2(G_{23} + K_{23})} \tag{5a}$$

$$u_2 = \frac{1}{8} E_{11} - \frac{G_{12}}{2} + \frac{(1 + 6v_{12} + v_{12}{}^2)G_{23}K_{23}}{2(G_{23} + K_{23})} \tag{5b}$$

with E_{11}, v_{12}, G_{12}, G_{23}, and K_{23} given by

$$E_{11} = v_f E_f + v_m E_m + \frac{4v_f v_m (v_f - v_m)^2}{\dfrac{v_m}{K_f + \dfrac{G_f}{3}} + \dfrac{v_f}{K_m + \dfrac{G_m}{3}} + \dfrac{1}{G_m}} \tag{6}$$

$$v_{12} = v_f v_f + v_m v_m + \frac{v_f v_m (v_f - v_m)\left[\dfrac{1}{K_m + \dfrac{G_m}{3}} - \dfrac{1}{K_f + \dfrac{G_f}{3}}\right]}{\dfrac{v_m}{K_f + \dfrac{G_f}{3}} + \dfrac{v_f}{K_m + \dfrac{G_m}{3}} + \dfrac{1}{G_m}} \tag{7}$$

$$G_{23} = G_m \left\{ 1 + \frac{v_f}{\dfrac{G_m}{G_f - G_m} + \dfrac{\left[K_m + \left(\dfrac{7}{3}\right)G_m\right]v_m}{2\left[K_m + \left(\dfrac{4}{3}\right)G_m\right]}} \right\} \tag{8}$$

$$G_{12} = G_m \frac{G_f(1 + v_f) + G_m v_m}{G_f v_m + G_m(1 + v_f)} \tag{9}$$

$$K_{23} = K_m + \frac{G_m}{3}$$

$$+ \cfrac{v_f}{\cfrac{1}{\left[K_f - K_m + \left(\frac{1}{3}\right)(G_f - G_m) \right]} + \cfrac{v_m}{\left[K_m + \left(\frac{4}{3}\right)G_m \right]}} \tag{10}$$

where the subscript f refers to fiber and the matrix volume fraction $v_m = 1 - v_f$.

The final composite density ρ_c can be expressed by

$$\rho_c = (\rho_p - \rho_r) v_p^* v_m + \rho_r v_m + \rho_f v_f \tag{11}$$

with v_p^* defined in Eq 2 and ρ_r, ρ_p, and ρ_f being the densities of the resin, filler, and fiber phases, respectively.

Volume-Weight Relationship

In the case of an unfilled SMC, the resin and fiber volume fractions, v_r and v_f, can be easily computed from the given fiber weight fraction, w_f, as follows:

$$v_r = \frac{\rho_f w_r}{\rho_f w_r + \rho_r w_f} \tag{12a}$$

$$v_f = 1 - v_r \tag{12b}$$

with the resin weight fraction $w_r = 1 - w_f$.

In the case of a filled SMC with constant resin volume, we must first compute the balancing resin and filler weight fraction, w_r and w_p, from given resin volume fraction and fiber weight fraction as

$$w_r = \frac{v_r \rho_r [(\rho_p - \rho_f)w_f + \rho_f]}{\rho_f [\rho_p - v_r(\rho_p - \rho_r)]} \tag{13a}$$

$$w_p = 1 - w_r - w_f \tag{13b}$$

Finally, the fiber and filler volume fractions can be computed as

$$v_f = \frac{\rho_r \, \rho_p \, w_f}{\rho_p \, \rho_f \, w_r + \rho_f \, \rho_r \, w_p + \rho_p \, \rho_r \, w_f} \tag{14a}$$

$$v_p = 1 - v_r - v_f \tag{14b}$$

Glass-Fiber Reinforced SMC

Since we are interested in achieving higher stiffness properties than in the conventional glass-fiber reinforced SMC composite, we will first examine the basic material properties of such a composite. As shown in Fig. 1, instead of using the commonly used stiffness parameter E_c/ρ_c, we have expressed the composite Young's modulus E_c and density ρ_c separately as functions of fiber weight fraction w_f. This is because the majority of car structure panels made of sheet form are governed either by bending stiffness or a combination of bending and membrane stiffnesses rather than membrane stiffness alone.

As expected, the higher the glass content, the stiffer the composite for both

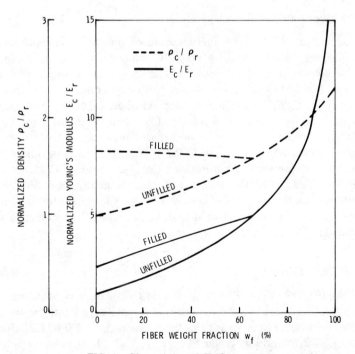

FIG. 1—*Glass-reinforced SMC composites.*

filled and unfilled cases. The rate of increase in filled cases, however, is fairly low. For example, an increase of glass weight by 67 percent from the commonly used SMC-30 to SMC-50 results in only a 21 percent increase in Young's modulus. The corresponding composite density, however, is slightly decreased (~ 2 percent). The highest E_c/E_r value in a commonly used filled glass-SMC (SMC-50) is 4.38.

In the case of unfilled SMC the rate of increase in stiffness with respect to fiber increase is much higher than in the filled counterpart. For example, SMC-50 is 55 percent stiffer and SMC-65 is 114 percent stiffer than SMC-30. Unlike the filled cases, densities of unfilled SMC increase with increasing fiber content. In this case, the density of SMC-50 weighs 15 percent and of SMC-65 weighs 31 percent more than that of SMC-30. The highest E_c/E_r value that can be achieved with the resin volume equal to 0.55 is 4.86 (SMC-65).

With the same glass content, the filled SMC is always stiffer and heavier than the unfilled one. For example, filled SMC-30 is 60 percent stiffer and 34 percent heavier than the unfilled SMC-30. With equal resin volume ($v_r = 0.55$), however, the filled SMC-30 is not only 34 percent less stiff but also 3 percent heavier than the unfilled counterpart (SMC-65).

Effect of Fiber Modulus

To evaluate the effect of the fiber modulus on the stiffness properties of the SMC composite, we will use the glass-SMC composition as a base to compose various hypothetical SMC composites, changing only Young's modulus of the fibers. The fiber modulus variation will range from $E_f/E_r = 20$ to 120 (for glass fiber, $E_f/E_r = 20.7$ and for carbon fiber, $E_f/E_r = 65$ to 138). The results are shown in Figs. 2 and 3 for unfilled and filled cases, respectively. The stiffness of the composite in both cases increases sharply when fiber modulus is increased. In the case $E_f/E_r = 120$, the stiffness ratio E_c/E_r of unfilled SMC-65 can reach 20, and this roughly corresponds to aluminum. In the case of filled SMC, the stiffness of the commonly used SMC-30 and SMC-50 can reach $E_c/E_r = 6.5$ and $E_c/E_r = 9.1$, respectively, with fiber modulus $E_f/E_r = 60$, and will go as high as 10.8 and 16.1, respectively, with fiber modulus $E_f/E_r = 120$.

Effect of Fiber Density

In evaluating the effect of fiber density on the stiffness of SMC composites we keep the properties of all the constituents constant except the fiber density. The variation of fiber density ranges from $\rho_f/\rho_r = 1.0$ to 2.25 (for glass fiber, $\rho_f/\rho_r = 2.26$ and for Kevlar fiber, $\rho_f/\rho_r = 1.29$).

For unfilled SMC the results are shown in Fig. 4, where both the normal-

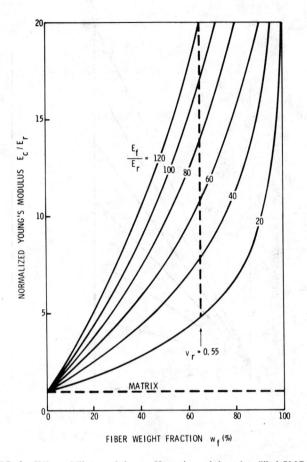

FIG. 2—*Effect of fiber modulus on Young's modulus of unfilled SMCs.*

ized composite modulus E_c/E_r and density ρ_c/ρ_r are expressed as functions of fiber content and fiber density. For the same fiber weight fraction, the lower the fiber density, the higher the composite stiffness and the lower the composite density. The maximum achievable stiffness for a given constant resin volume, however, remains almost constant. For example, the unfilled SMC-40 has $E_c/E_r = 2.85$ and $\rho_c/\rho_r = 1.60$ if $\rho_f/\rho_r = 2.25$, and has $E_c/E_r = 3.89$ and $\rho_c/\rho_r = 1.09$ if $\rho_f/\rho_r = 1.25$. On the other hand, the maximum achievable stiffness ratio is $E_c/E_r = 4.86$ for $v_r = 0.55$. This level of stiffness can be obtained with SMC-65 if $\rho_f/\rho_r = 2.25$, SMC-55 if $\rho_f/\rho_r = 1.5$, and SMC-45 if $\rho_f/\rho_r = 1.0$. The densities ρ_c/ρ_r of these three compositions, however, are different and are equal to 1.56, 1.22, and 1.0, respectively.

For the filled cases (Fig. 5), similar trends can be observed as for the un-

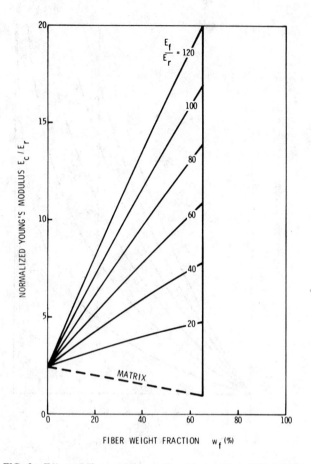

FIG. 3—*Effect of fiber modulus on Young's modulus of filled SMCs.*

filled cases, except that the effect of changing fiber density on the composite stiffness is smaller than on the composite density. For instance, a filled SMC-30 has $E_c/E_r = 3.64$ and $\rho_c/\rho_r = 1.62$ if $\rho_f/\rho_r = 2.25$, and has $E_c/E_r = 3.99$ and $\rho_c/\rho_r = 1.39$ if $\rho_f/\rho_r = 1.25$.

Effect of Fiber Type

In general, the fiber modulus and density cannot be varied independently. In this section, comparisons are made among SMC-composites reinforced by four commercially available fiber systems: glass, Kevlar, carbon, and high-modulus carbon. The material properties of these fibers along with resin and filler properties are illustrated in Fig. 6. It is interesting to note that all other types of fibers considered are not only stiffer but also lighter than the glass fi-

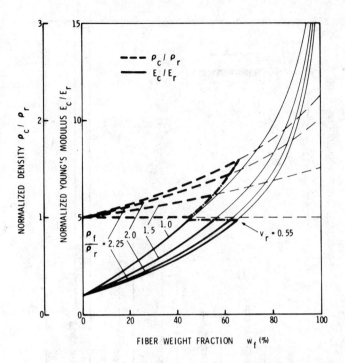

FIG. 4—*Effect of fiber density on Young's modulus and density of unfilled SMCs.*

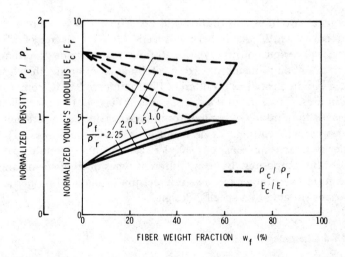

FIG. 5—*Effect of fiber density on Young's modulus and density of filled SMCs.*

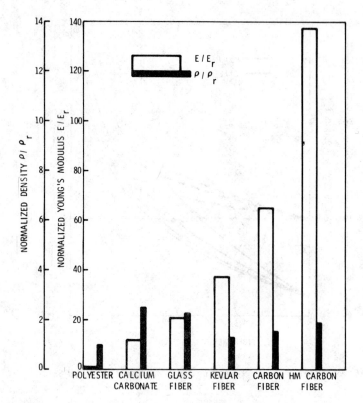

FIG. 6—*Comparison of constituent properties of SMC composites.*

bers. Consequently, any fiber substitution will result in an SMC composite that is more effective than the glass fiber reinforced SMC. The degree of efficiency, however, will depend on the type of stiffness requirement governing the design, such as bending, membrane, or a combination of both. The elastic moduli (E_c, ν_c) and density (ρ_c) of different fiber reinforced SMC composites are shown in Figs. 7 to 9 for unfilled cases and in Figs. 10 to 12 for filled cases. As expected, the ranking of composites in order of decreasing stiffness is: high-modulus carbon, carbon, Kevlar, and glass SMC composites. The ranking in order of increasing density is: Kevlar, carbon, high-modulus carbon, and glass SMC composites. By using different types of fibers or mixing them in hybrid form, we can achieve a wide spectrum of material properties of SMC composites suitable to a specific design.

Summary and Conclusions

Using the analytical method of Ref *2* we have evaluated the effect of fiber systems on the stiffness properties of SMC composites. Analytical results are

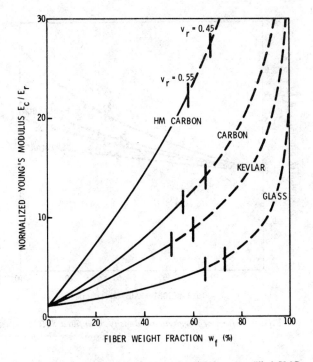

FIG. 7—*Effect of fiber type on Young's modulus of unfilled SMCs.*

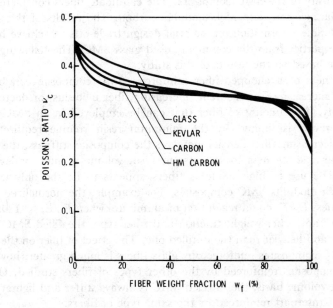

FIG. 8—*Effect of fiber type on Poisson's ratio of unfilled SMCs.*

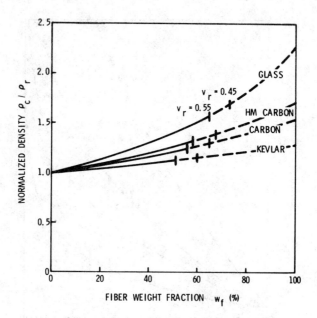

FIG. 9—*Effect of fiber type on density of unfilled SMCs.*

presented to illustrate the influences resulting from changing fiber content, fiber modulus, fiber density, and fiber type on the effective stiffness properties and density of the SMC composite. The candidate fibers considered are glass, Kevlar, carbon, and high-modulus carbon. The results of this study provide valuable information on material design trade-offs to achieve higher stiffness properties from the commonly used glass SMC. The following conclusions are based on the results of this study.

The stiffness of a chopped-fiber reinforced SMC composite can be increased by increasing fiber content, increasing fiber modulus, or decreasing fiber density. The amount of fiber that can be employed in an SMC composite, however, is limited by the minimum resin volume requirement. Although decreasing fiber density increases the composite stiffness, the maximum achievable stiffness for a constant resin volume is more or less unchanged. The use of high modulus fibers appears to be the only way to achieve high-modulus SMC composites. For example, the normalized composite stiffness E_c/E_r could reach that of aluminum when $E_f/E_r = 120$.

With the same fiber weight fraction and fiber type, the filled SMC is always stiffer and heavier than the unfilled one. The effect of filler on the stiffness of SMC composites reinforced by glass fibers is much greater, however, than on composites reinforced by the other types of fibers studied. On an equal resin volume basis, the unfilled SMC is always stiffer and lighter than the filled counterpart reinforced by the same type of fibers.

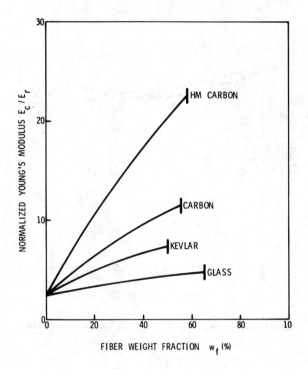

FIG. 10—*Effect of fiber type on Young's modulus of filled SMCs.*

Among the SMC composites considered in this study, the ranking by order of decreasing stiffness is: high-modulus carbon, carbon, Kevlar, and glass. In the case of the commonly used unfilled SMC-65, the corresponding Young's modulus ratios are 5.35:2.89:2.01:1.0, with the corresponding density ratios being 0.87:0.82:0.74:1.0. For the case of the commonly used filled SMC-30, the corresponding Young's modulus ratios are 3.87:2.23:1.60:1.0, with the corresponding density ratios being 0.90:0.87:0.80:1.0.

The analytical method used in computing the effective stiffness properties of chopped-fiber reinforced SMC composites has shown good agreement with experimental results for glass-fiber reinforced SMC composites [2]. However, in applying the proposed method to calculate the effective stiffness of other types of chopped-fiber reinforced SMC composites, one must keep in mind that the imposed assumptions discussed earlier represent idealized conditions for the composite. In reality, these assumptions are simply approximations. The facts that the bonding might not be perfect, the fillers might be agglomerated, the fibers might not be uniformly distributed, and that there might be voids all suggest that the estimated results according to the proposed method could be higher than experimentally measured values.

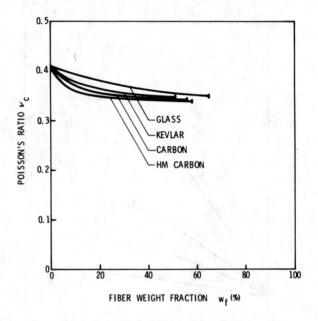

FIG. 11—*Effect of fiber type on Poisson's ratio of filled SMCs.*

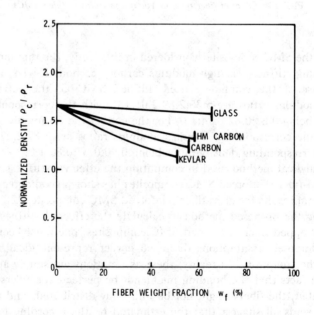

FIG. 12—*Effect of fiber type on density of filled SMCs.*

References

[1] Chang, D. C., "Optimal Designs with Alternate Materials Suitable for High-Volume Production," in *Proceedings*, Third International Conference on Vehicle Structural Mechanics, Detroit, Oct. 1979, pp. 27-33.
[2] Chang, D. C. and Weng, G. J., *Journal of Materials Science*, Vol. 14, Sept. 1979, pp. 2183-2190.
[3] Jutte, R. B., "Structural SMC—Material, Process, and Performance Review," Paper 780355, presented at SAE Automotive Engineering Congress, Society of Automotive Engineers, Detroit, Feb. 1978.
[4] Hashin, Z., *Journal of Applied Mechanics*, Vol. 29, 1962, p. 143.
[5] Christensen, R. M. and Waals F. M., *Journal of Composite Materials*, Vol. 6, 1972, p. 518.

R. W. Tung[1]

Effect of Processing Variables on the Mechanical and Thermal Properties of Sheet Molding Compound

REFERENCE: Tung, R. W., "**Effect of Processing Variables on the Mechanical and Thermal Properties of Sheet Molding Compound,**" *Short Fiber Reinforced Composite Materials, ASTM STP 772,* B. A. Sanders, Ed., American Society for Testing and Materials, 1982, pp. 51-63.

ABSTRACT: The increasing usage of sheet molding compound (SMC) materials in the automotive industry for weight reduction motivated this study on the effect of processing on SMC properties. The processing variables investigated were cure time, mold temperature, and mold pressure. A total of 15 different molding conditions were used to produce flat SMC panels from a laboratory mold. Tensile, flexural, and impact properties and heat deflection temperatures of these sample panels were measured. It was found that while the processing variables studied have little effect on the mechanical properties, they have a more severe effect on the thermal property.

KEY WORDS: sheet molding compound, processing variables, mechanical properties, thermal properties, tensile strength, tensile modulus, percent elongation, flexural strength, flexural modulus, impact strength, heat deflection temperature, cure time, cure temperature, cure pressure, compression molding, automotive materials, formulation, dynamic mechanical properties

Sheet molding compound (SMC) materials, because of their high strength and low weight characteristics, have been used in recent years for sheet metal replacement in the automotive industry. These materials are usually composed of polymeric-based neat resins, such as polyester, and reinforced with chopped fibers of fiberglass. A high percentage of inert fillers (for example, calcium carbonate) is also used in some automotive grades of SMC to reduce material cost.

[1] Senior Project Engineer, General Motors Manufacturing Development, General Motors Technical Center, Warren, Mich. 48090.

Numerous studies have been made on the mechanical and physical properties of SMC materials, among which may be cited work by Heimbuch and Sanders [1],[2] Denton [2], Jutte [3], and Enos et al [4]. In addition, processing techniques for SMC materials have been studied by Smith and Suh [5], Burns [6], Kobayashi and Pelton [7], and Marker and Ford [8]. However, when questions are raised on the relationships between processing and properties, very little information can be found in the literature to delineate such relationships. Among this limited work, a recent study by Jones [9] investigated the effect of flow front velocity and charge pattern on the tensile properties of SMC panels; another study by Mallick and Raghupathi [10] investigated the effect of cure, preheating, and postcooling on the mechanical properties of thick SMC panels.

The purpose of this work was to investigate the effect of processing variables on the mechanical and thermal properties of SMC materials. The processing variables included in this study were cure time, mold temperature, and mold pressure. A total of 15 different molding conditions were used to include variations of 121 to 177°C in mold temperature, 2.83 to 10.3 MPa in mold pressure, and 0.5 to 10 min in cure time. Mechanical properties tested using conventional ASTM methods were tensile, flexural, and impact properties. The thermal property measured was the heat deflection temperature determined by the thermal mechanical analysis technique.

The final goal of this study was to establish a relationship between the processing variables and the mechanical and thermal properties of a molded SMC panel. With this relationship established, a reasonable prediction of the properties of a molded component with known processing conditions can be made. Therefore this type of information can be used to optimize the processing conditions of SMC to achieve the best processing efficiency and product performance.

Sample Preparation

SMC Formulation

The SMC material used was SMC-R25, a polyester resin based system reinforced with 25 percent by weight chopped E-glass. The formulation of this material is shown in Table 1.

SMC Molding Conditions

In the molding of a sheet molding compound, many processing parameters can usually be changed to tailor the desirable properties in a molded part. These parameters include cure time, mold temperature, mold pressure,

---

[2]The italic numbers in brackets refer to the list of references appended to this paper.

TABLE 1—*SMC-R25 formulation.*

Ingredient	Weight Percent
Resin	29.4
Filler	41.8
Catalyst	0.3
Thickener	1.5
Internal release	1.1
2.54-cm chopped E-glass	25.0

material flow front velocity, parallelism of the mold, and charge pattern of the material. Three parameters—cure time, mold temperature, and mold pressure—were chosen for study since they can be easily changed.

The cure time is defined as the time during which pressure was actually applied on the material. Mold temperature was measured by thermocouples at the mold surface, and the mold pressure was calculated from the press ram force and the panel surface area. During panel molding, these three parameters were varied by holding two parameters constant during each series of molding and systematically changing the third parameter. A complete spectrum of processing conditions was covered by this method and 15 panels molded at different conditions were obtained.

The cure time, mold temperature, and mold pressure covered were in the range of 0.5 to 10 min, 121 to 177°C, and 2.8 to 10.3 MPa. Flat panels 53 by 61 by 0.3 cm were molded in a 500-ton laboratory press under closely controlled conditions. The mold was charged with approximately 90 percent of its surface area in all the moldings. Test specimens were cut from panels in random directions to minimize any possible directional effect that could be reflected in the data. The specimens were conditioned at 24°C and 50 percent relative humidity for a minimum of 48 h before testing.

Experimental Techniques

Mechanical Properties

The mechanical properties investigated were tensile, flexural, and impact properties. Tensile strength, modulus, and elongation at break were tested at room temperature in accordance with ASTM Test for Tensile Properties of Plastics (D 638), and a crosshead speed of 5.1 mm/min was used. The flexural strength and modulus of the composite were determined by ASTM Test for Flexural Properties of Plastics and Electrical Insulating Materials (D 790), with a crosshead speed of 1.3 mm/min. In Izod impact strength was determined on notched specimens by ASTM Tests for Impact Resistance of Plastics and Electrical Insulating Materials (D 256). Five specimens from the same plaque were tested to obtain the average and the standard deviation for all three mechanical properties tested. The average and one standard deviation were used in plotting Figs. 1 to 9.

Thermal Property

The thermal property of SMC-R25 investigated was the heat deflection temperature (HDT). Because of the reinforcing fibers, the conventional method for the determination of heat deflection temperatures, ASTM Test for Deflection Temperature of Plastics Under Flexural Load (D 648), was not applicable to SMC-R25. Instead, the HDT was determined by the thermal mechanical analysis technique in the penetration mode. A DuPont 943 thermal mechanical analyzer (TMA) was used throughout this study. In the thermal mechanical analyzer, a quartz probe with a flat cylindrical tip was placed on top of a composite specimen approximately 0.5 cm square, then the temperature of the specimen was raised at a rate of 10°C/min. The displacement of the quartz probe due to penetration of the specimen surface was then accurately measured as a function of temperature. A loading of 0.2 N (20 g) was used on top of the specimen to achieve the desirable penetration throughout this study.

Since the base resin in SMC-R25 is a polymeric thermoset material, the behavior of SMC-R25 in TMA penetration mode was found very similar to that of a typical thermoset polymer. The only exception was the larger data scattering that was observed in the SMC material due to the presence of fibers and their often nonuniform distributions in the specimen surface. The

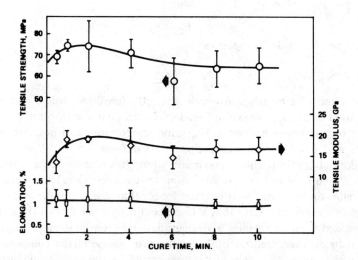

FIG. 1—*Effect of cure time on tensile properties.*

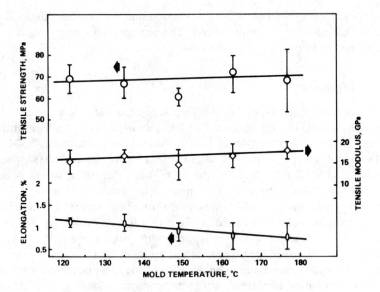

FIG. 2—*Effect of mold temperature on tensile properties.*

peak temperature of the differential curve obtained in the TMA, which indicates the temperature of maximum penetration rate, was arbitrarily defined as the heat deflection temperature.

Five specimens molded under the same conditions were tested in the TMA, and the averages and the standard deviations of data were calculated and plotted in Figs. 10 to 12.

Effects on Mechanical Properties

Tensile Properties

The effect of cure time on tensile strength, tensile modulus, and percent elongation at break is shown in Fig. 1. All data points in this graph were obtained on specimens molded at the same temperature and pressure, namely 149°C and 6.9 MPa, but cure times varied from 0.5 to 10 min. Tensile strength data indicate that a maximum strength is obtained at a cure time of 2 min. At cure times shorter than 2 min, tensile strength is lower because the material is undercured. At cure times longer than 2 min, tensile strength also decreases to such an extent that a 15 percent drop is seen when the cure time is increased from 2 to 8 min. This degradation in tensile strength is probably caused by an overcured condition in the resin phase of the composite. It is known that excessive localized heat generated by the exothermic cure reaction can lead to a partial breakdown of the cross-linked structure in the ther-

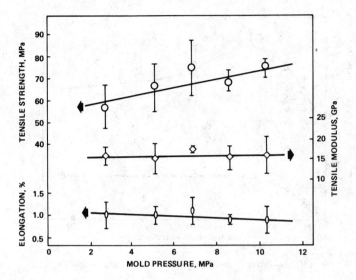

FIG. 3—*Effect of mold pressure on tensile properties.*

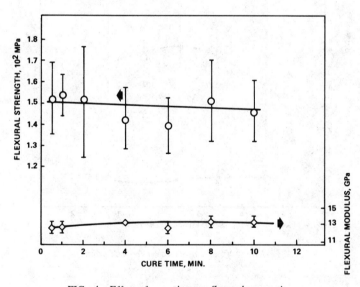

FIG. 4—*Effect of cure time on flexural properties.*

moset. A similar type of structural breakdown and degradation in tensile properties has been reported in the curing studies of rubber [*11*], also a thermoset material. This similarity in behavior is not surprising if we recognize the fact that both SMCs and elastomers are polymeric-based thermosets. In a

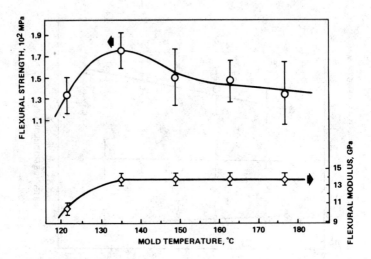

FIG. 5—*Effect of mold temperature on flexural properties.*

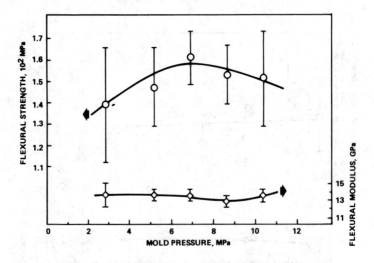

FIG. 6—*Effect of mold pressure on flexural properties.*

later section, we will see that the thermal property of SMC-R25 is also degraded at prolonged curing times.

The effect of cure time on tensile modulus (center curve in Fig. 1) follows the same trend as for tensile strength. A small decrease in the tensile modulus is seen at prolonged cure times. The percent elongation at break, indicated by the bottom curve in Fig. 1, was hardly affected by the cure time.

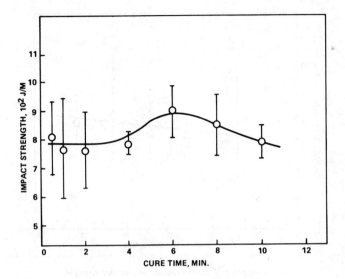

FIG. 7—*Effect of cure time on impact strength.*

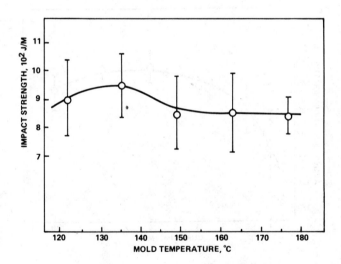

FIG. 8—*Effect of mold temperature on impact strength.*

The effect of mold temperature on the tensile properties can be seen in Fig. 2. All specimens tested in the study of temperature effect were molded with 2-min cure time and 6.9 MPa mold pressure with the mold temperature varied between 121 and 177°C.

A small increase in both tensile strength and modulus with increasing

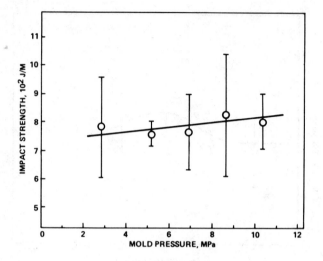

FIG. 9—*Effect of mold pressure on impact strength.*

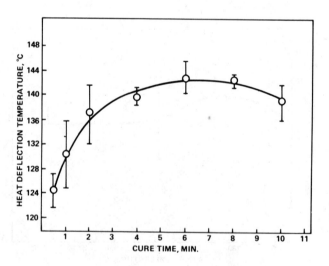

FIG.10—*Effect of cure time on HDT.*

mold temperature is observed. This increase can be attributed to the faster cure rate in the resin phase at higher mold temperatures. Therefore a panel molded at higher temperatures will have a better cured structure than one molded at lower temperatures. For a thermoset polymeric material, a better cured structure implies a structure of higher cross-linking density, less void

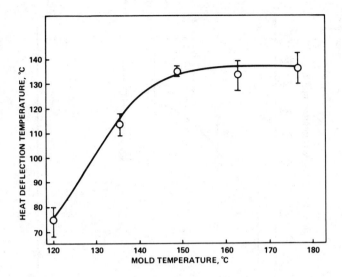

FIG. 11—*Effect of mold temperature on HDT.*

content, and consequently higher mechanical strength. The percent elongation data (bottom curve in Fig. 2) indicated a drop of this property in panels molded at higher temperatures. This is expected since at higher mold temperatures, higher modulus and thus stiffer materials were produced.

The effect of mold pressure on tensile properties is illustrated in Fig. 3. Specimens used in this pressure effect study were molded at constant temperature and time, (149°C and 2 min), while the mold pressure was changed from 2.8 to 10.3 MPa. Mold pressure was found to have a significant effect on the tensile strength of SMC-R25; a 30 percent increase is seen when the mold pressure is increased from 2.8 to 10.3 MPa. The tensile modulus, on the other hand, was only increased modestly. The large increase in tensile strength with increasing mold pressure can be explained if the bond between fibers and resin phase is considered. At high mold pressures, more resin is forced into fiber bundles to fill the voids, which leads to a more effective bonding between resin and fibers. This more effective bonding will in turn make the stress transfer from resin to fibers more efficient, and thus the composite has a higher total strength. Also, since void content is considered a defect, smaller void content in the panels molded at high pressures also contributes to the higher strength of those panels. In a recent study by Burroughs and Blaine [12], a positive effect of mold pressure on the dynamic mechanical properties of graphite-reinforced composite was also found. The effect of mold pressure on the percent elongation (bottom curve in Fig. 3) is small and insignificant.

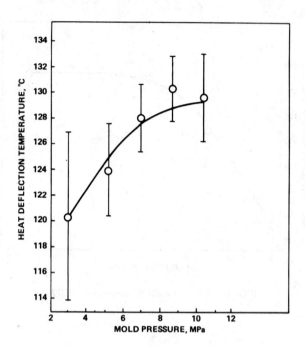

FIG. 12—*Effect of mold pressure on HDT.*

Flexural Properties

The effect of cure time on flexural strength and modulus was studied (Fig. 4). Because of large data scattering, no distinct trend can be seen in the dependence of strength and modulus on the cure time. The only observation that can be made is that the maximum flexural strength was achieved at relatively short cure times.

A more distinct effect of mold temperature on flexural properties can be seen in Fig. 5. At low mold temperature (for example, 122°C) both flexural strength and modulus are low because the material is undercured. As mold temperature increases, a peak strength and modulus value are reached at 135°C. At still higher mold temperatures, flexural strength decreases slightly while flexural modulus remains unchanged.

A similar effect is seen in the pressure dependence of flexural properties (Fig. 6) even though considerable data scattering was found in the strength data. This scattering is more pronounced in panels molded at low pressures, which indicates that panels of less consistent quality were obtained at low pressures. It appears that flexural strength increases with mold pressure until a peak value is reached at 6.9 MPa. Further increases of mold pressure cause a small drop in the flexural strength. The stiffness of the composite, as indicated by the modulus values, is not affected by the mold pressure.

Impact Property

Another mechanical property investigated was the impact strength of SMC-R25. Figure 7 shows the effect of cure time on the impact strength of notched specimens. At a cure time of 6 min, the maximum impact strength was obtained for the composite. A lower impact strength was obtained when the cure time was longer or shorter than 6 min, apparently caused by over-cured or undercured condition.

A similar effect is seen on the impact strength by variations in the mold temperature (Fig. 8). Again, there is an optimum mold temperature (135°C) at which the peak impact strength was obtained. A deterioration in impact strength is seen when panels were molded at either higher or lower mold temperatures.

A positive effect of mold pressure on the impact strength is seen in Fig. 9. This is quite similar to the previously discussed effect of mold pressure on tensile and flexural properties (Figs. 3 and 6). It is assumed that the same mechanism discussed previously caused this improvement in properties.

Effects on Thermal Property

The thermal property investigated in this study was the heat deflection temperature by the TMA method. Information on the heat deflection temperature of an SMC material is important from both the design and processing point of view. Firstly, HDT is an important design parameter, since most mechanical properties of SMC suffer a drastic reduction at this temperature. For instance, it has been shown that the tensile strength of SMC-R25 decreases by as much as 50 percent at test temperatures higher than HDT of the material [13]. Secondly, HDT is a very sensitive function of processing parameters. Information on this relationship is important to a material processor in order to obtain the optimum HDT in a molded SMC part.

The HTD of SMC-R25 usually exhibited smaller data scattering than the large scattering in the mechanical test data. Figure 10 shows the effect of cure time on the heat deflection temperatures of SMC-R25. When the cure time was less than 2 min, very low values of HDT were obtained because this combination of time and temperature was insufficient for a complete cure. A more gradual increase in HDT then takes over at cure times longer than 2 min until a peak temperature is reached at 6 min cure. At even longer cure times, a decrease in HDT is seen that signals a possible degradation in the material due to prolonged curing. Similar deterioration in tensile properties at prolonged curing was discussed previously (Fig. 1). It is believed that both property deteriorations were caused by the same type of structural degradation in SMC-R25 due to prolonged heating.

Mold temperature was also found to have a severe effect on the heat deflection temperatures of SMC-R25. This result was not unexpected since the rate of cure, or cross-linking, is a sensitive function of temperature. In Fig. 11, it is shown that the HDT of the composite was almost doubled when the mold temperature was increased from 120 to 148°C. Once this maximum HDT is reached at 148°C mold temperature, it remains constant for further increases in mold temperature.

Mold pressure did not have as strong an effect on HDT as did cure time and mold temperature; however, a distinct positive effect of mold pressure on the HDT can be seen in Fig. 12. When the mold pressure was increased from 2.8 to 10.3 MPa, an increase of 10°C in HDT was achieved. Again, at low pressures, a large scattering of data is observed due to the inconsistency in quality of the panels molded. It is also noticed in Fig. 12 that HDT increases more rapidly at low mold pressures than at high mold pressures. For instance, approximately 80 percent of the increase in HDT occurred in the pressure range of 2.8 to 6.9 MPa. This type of behavior, where the thermal property reaches a plateau region and then remains almost constant, is consistently observed in the study of all three processing variables.

Summary and Conclusions

The effect of processing variables on the mechanical and thermal properties of sheet molding compound material was investigated. Fifteen different molding conditions involving various cure times, mold temperatures, and mold pressures were used to produce the specimens. Tensile, flexural, and impact properties and heat deflection temperatures of these specimens were then measured. It was found that while processing variables have little effect on the mechanical properties, they have a more severe effect on the thermal property. These findings can be summarized as follows:

1. A definite trend of property variations on the processing variables can be established on almost all properties investigated.
2. Data scattering in flexural properties was sometimes too large to make a reasonable deduction of the trend of property change, especially in specimens molded at low pressures.
3. Thermal property data have higher reproducibility than mechanical property data.
4. Prolonged cure time was found to have negative effects on both mechanical and thermal properties of SMC, a finding consistent with that for other types of thermoset polymeric materials.
5. Mold pressure was found to have a significant effect on both mechanical and thermal properties. A similar effect was found by other investigators in the study of dynamic mechanical properties of composites.

6. If we assume that parts of similar thickness molded in a production press have the same properties as a flat panel molded in a laboratory press, data obtained in this study can be used to optimize molding conditions of SMC-R25 material.

Future studies can be made on the effect of other processing variables, such as the velocity of mold closing, percent charge in the mold, different charge patterns, and the effect of those variables on properties of parts with thick sections. In thick sections we expect the effect of heat transfer to have a significant contribution to the property variations.

Acknowledgments

The author would like to thank the Plastics Processing Development Department personnel at General Motors Manufacturing Development for their expert molding of the flat panels used in this study. The laboratory assistance of Mr. Thomas Bamford in obtaining the TMA data and the critical comments of Mr. William Todd on the manuscript are also greatly appreciated.

References

[1] Heimbuch, R. A. and Sanders, B. A., "Mechanical Properties of Automotive Chopped Fiber Reinforced Plastics," Report MD-78-032, General Motors Corp., Warren, Mich., 1978.
[2] Denton, D. L., "Mechanical Properties of an SMC-R50 Composite," Owens-Corning Fiberglas Corp., 1979.
[3] Jutte, R. B., Paper 780355, Society of Automotive Engineers, 1978.
[4] Enos, J. H., Erratt, R. L., Francis, E., and Thomas, R. E., "Structural Performance of Vinyl Ester Resin Compression Molded High Strength Composites," in *Proceedings*, 34th SPI Conference, New Orleans, Section 11-E, 1979.
[5] Smith, K. L. and Such, N. P., *Polymer Engineering and Science*, Vol. 19, No. 12, Sept. 1979, p. 829.
[6] Burns, R., *Polymer-Plastics Technology and Engineering*, Vol. 10, No. 2, 1978, p. 165.
[7] Kobayashi, G. S. and Pelton, E. R., "Process Development for Fabricating Sculptured Decorative Interior Aircraft Panels Using Sheet Molding Compounds," paper presented at 32nd SPI Annual Technical Conference, Section 2-A, 1977.
[8] Marker, L. and Ford, B., "Rheology and Molding Characteristics of Glass Fiber Reinforced Sheet Molding Compound," paper presented at 32nd SPI Annual Technical Conference, Section 16-E, 1977
[9] Jones, J. J., unpublished results, General Motors Manufacturing Development, Warren, Mich., 1979.
[10] Mallick, P. K. and Raghupathi, N., *Polymer Engineering and Science*, Vol. 19, No. 11, Aug. 1979, p. 774.
[11] Morton, M., Ed., *Rubber Technology*, 2nd ed., Van Nostrand Reinhold, New York, Chapter 4, 1973.
[12] Burroughs, P. and Blaine, R., paper presented at 9th North American Thermal Analysis Society Conference, Chicago, Sept. 1979.
[13] Gray, A. P., "Thermal Analysis Application Study No. 2," Perkin Elmer Corp., Sept. 1972.

M. J. Owen[1]

Static and Fatigue Strength of Glass Chopped Strand Mat/Polyester Resin Laminates

REFERENCE: Owen, M. J., "Static and Fatigue Strength of Glass Chopped Strand Mat/Polyester Resin Laminates," *Short Fiber Reinforced Composite Materials, ASTM STP 772*, B. A. Sanders, Ed., American Society for Testing and Materials, 1982, pp. 64–84.

ABSTRACT: Chopped strand mat/polyester resin laminates are regarded as plane isotropic. However, under load the aligned fibers provide reinforcement, whereas the transverse fibers initiate progressive damage by filament debonding and resin cracking. This transverse damage rapidly involves the aligned fibers, particularly under repeated loading. Static and fatigue strength data have been developed for recognizable states of damage as well as final separation of the specimens. The effect of mean load on stress amplitude has been studied, together with cumulative damage and the effect of stress concentrators. Failure theories for biaxial stress conditions have been examined using thin-walled tubes. The biaxial tensile condition is particularly damaging.

There appear to be three main difficulties in applying conventional data to design. The *S-N* curves appear to be straight lines and there is difficulty in extrapolating to long lives; there is a marked adverse size effect on strength; and designing to avoid incipient damage at stress concentrators gives very uncompetitive designs. Fracture toughness and macroscopic crack growth studies appear to offer an alternative approach to the prediction of strength.

KEY WORKS: composite materials, glass fiber, chopped strand mat, polyester, fatigue, Goodman diagram, scatter, fracture toughness, crack propagation, damage stress rupture

Research and development concerned with the properties and applications of composite materials has proceeded under a variety of driving influences for over 30 years. Funding has been mainly derived from defense or aerospace sources and the emphasis has been on high-performance materials. Materials producers have an interest in selling the maximum quantities of fiber and resin. Major markets have been developed in chopped glass fibers and unsaturated polyester resins. This combination has been traditionally

[1]Department of Mechanical Engineering, University of Nottingham, U.K.

regarded as a low-cost, low-performance material and has attracted only a tiny proportion of the research budgets available. Prompted by the British Plastics Federation (BPF) in 1962, the author has undertaken a number of studies on glass chopped strand mat (CSM) and polyester resin (PR) combinations. Although there have been many publications, these are now widely scattered and it is timely to draw them together into a coherent statement.

Design procedures for fiber reinforced plastics (FRP) were slow to emerge and for a long time it was implicit that the procedures developed for metals would apply to FRP. High-performance FRP generally involves various configurations of well-collimated fibers. The mechanics of such materials have been collated and design procedures continue to be developed and proved. Short fiber reinforcements are usually assumed to be randomly distributed in one plane and are generally regarded as forming a plane isotropic composite. As such they seem to form a distinct class of materials lying somewhere between the isotropic metals and the laminated orthotropic high performance composites.

The original request from the BPF was to develop data in the form of conventional fatigue and creep curves for CSM/PR laminates. The only substantial work on fatigue and creep of glass reinforced plastics (GRP) published before 1962 was the work of Boller and his associates at the U.S. Forest Products Laboratory. Almost exclusively, that work was on glass fabric reinforced polyester and epoxy resins with specimens cut so that their axes were parallel to a principal material axis, and the criterion of failure was separation of the specimens. No attention had been paid to scatter or to damage processes. Since that time the author and his associates have studied the static and fatigue strength of chopped strand mat polyester resin composites under both uniaxial and biaxial loading, and have considered failure processes, some aspects of cumulative damage, macroscopic crack propagation, and fracture mechanics.

Failure Mechanisms

CSM consists of 5-cm glass multifilament strands randomly arranged in a plane and bound together with a resin-soluble binder. There is considerable variation in superficial density. When CSM has been incorporated with PR by the wet lay-up technique there is usually a significant number of spherical voids. These are readily removed from thin sheets, but are progressively more difficult to remove from thick laminates consisting of several layers of mat. The CSM/PR laminates are generally translucent and can be observed during loading with an optical microscope. General purpose polyester resins are usually fairly brittle; as previously reported [1-6] the first signs of damage in CSM/PR occur at about 30 percent of the ultimate tensile strength.[2] Macro-

---

[2]The italic numbers in brackets refer to the list of references appended to this paper.

scopically there is a slight whitening of the laminate, usually only noticeable against a dark background. At this stage, using a thin laminate and transmitted light, the microscope can detect strands lying normal to the loading axis that are revealed as closely spaced black lines. If a polished cut edge is examined two features are noticeable; there are groups of closely spaced fibers (strand groups) isolated in fields in resin; among fibers that are normal to the cut surface (that is, perpendicular to the line of load) it will be seen that some fibers have separated from the resin matrix. Usually these "debonds" extend across the width of the strand group. It is these debonds that give rise to the closely spaced black lines in the transmitted light picture. If the loading is increased, the debonding intensifies and starts to affect fibers that are at angles other than 90 deg to the line of load. Depending on the resin formulation, at approximately 70 percent of the ultimate tensile strength (UTS) some of the debonds extend out into the resin-rich areas as resin cracks that can reach the specimen surface. Similar damage occurs under static loading, long-term creep loading, and fatigue loading. Although spherical voids are often present in CSM/PR, they do not seem to affect failure. In cut sections resin cracks sometimes pass close to voids without intersecting them.

It is possible to quantify the development of debonding and resin cracking. Owen and Howe [5,6] obtained micrographs of sites on the polished cut edge of a specimen at various fractions of the ultimate load, or at various fractions of fatigue life. They then counted the number of debonded fibers or measured the length of the resin cracks in each successive photograph. To deal with the different topography of the sites chosen, they normalized the results by dividing by the amount of damage present at failure. The development of damage was nonlinear with respect to load or life, and both debonding and resin cracking were much more intensively developed under fatigue loading than under static loading. Another feature of fatigue loading sometimes seen under the microscope is that at the later stages of the fatigue life there is fine debris both in the debond region and in the resin cracking regions. Scanning electron micrographs of fatigue fracture surfaces usually reveal this fine debris, which is not present on static failures.

It is possible to suppress resin cracking under static loading by variations in resin formulation. For example, Owen and Rose [4] added a flexibilizing agent to a general purpose orthophthalic polyester resin. Above 10 percent addition this had a marked effect on the strain to failure of a resin cast but not on the CSM/PR laminates (Fig. 1). The addition of about 15 percent flexibilizer suppresses resin cracking entirely in a static test and significantly increases the UTS (Fig. 2). It also reduces the modulus of the resin and begins to reduce the modulus of the laminate. It does not suppress debonding, which occurs at the same strain and correspondingly at a slightly lower stress (Fig. 2). However, if fatigue tests are carried out it is found that the resin cracking reappears after about 1000 cycles; by about 10^6 cycles the flex-

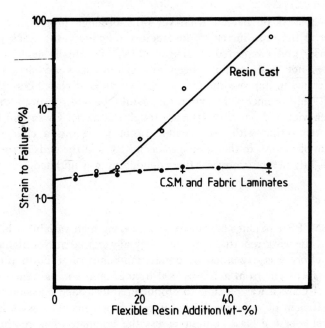

FIG. 1—*Change in strain to failure produced by adding polypropylene maleate adipate flex-ibilizer to an orthophthalic polyester resin* [4].

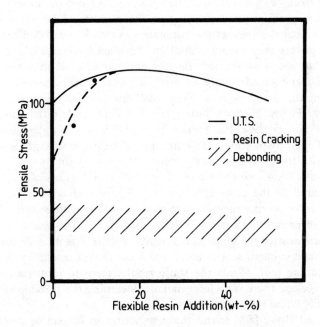

FIG. 2—*Effect of flexible resin addition on strength of CSM/PR laminate* [4].

ible resin conveys no benefit whatever [4]. Although progressively adding flexibilizer to the resin increases the fracture toughness, the crack propagation behavior under repeated loading conditions is actually made worse [7]. There does not seem to have been any systematic work carried out on polyester resins to improve the fatigue performance of glass fiber reinforced laminates. Improvements in the resin should be judged from the crack growth behavior and not from fracture toughness alone. Owen and Rose [4] also discovered that with very flexible resins there was a change in the mechanism of failure, in that damage was initiated at the ends of the aligned strands slightly before transverse fiber debonding was initiated.

Scatter

The CSM/PR laminates are regarded as having high variability in properties. When the work reported here was originally started at Nottingham, considerable effort was expended to obtain laminates of uniform properties. With specimens cut from a 300 by 300 by 6.25 mm laminate incorporating six layers of CSM it was possible to produce parallel-sided tension specimens whose coefficient of variation did not exceed 5 percent. The essential factor was control over the glass content across the laminate. The reinforcement was usually found to have a preferred orientation along the axis of the roll and it was necessary to lay up alternate layers at right angles and to reject unusually thick or thin sections. The resin was weighed out exactly and subdivided into portions corresponding to the number of layers of reinforcement. The final thickness of the laminate was closely controlled by laying up in a steel picture frame mold. Variability between laminates is largely due to variations in the reinforcement density and variation of thickness between laminates. Another advantage of using a picture frame mold is to control flow during the lay-up process. Even small amounts of flow produce marked anisotropy. Figure 3 shows histograms for UTS and initial modulus [8]. Figures 4 and 5 show the variability of UTS and ultimate compressive strength (UCS) by plotting a large number of results of strength against glass content with the glass content determined for each individual specimen [9]. These results show also the effect of two different batches of reinforcement manufactured to the same specification. Figure 6 shows the variability of fatigue lives for 30 specimens at each of three different stress levels plotted in the form of probability against cycles to failure [3]. The scatter is no worse than typical scatter for metal components. Figure 7 is an S-N curve which shows a static strength scatter band and a number of results for 0 to tension fatigue loading [10]. While the static results seem to form part of a continuous envelope there is no information concerning the possible validity of a strength life equal rank assumption.

Owen and Howe [5,6] found that the scatter in fatigue or stress rupture

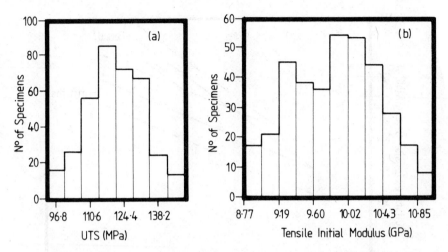

FIG. 3—(a) *Distribution of UTS for CSM/PR specimens* [8]. (b) *Distribution of initial modulus for CSM/PR specimens* [8].

properties could be reduced by dividing the test stress for each specimen by the experimentally determined UTS of its neighbor cut from the laminate.

Uniaxial Fatigue and Stress Rupture

Owen and Smith [10] carried out extensive fatigue tests on CSM/PR laminates fabricated with two orthophthalic resins. They presented their results as conventional S-N curves which included the static strength scatter band and six replicate fatigue tests at each of four stress levels plus isolated tests at other stress levels (Fig. 7). They considered the effect of mean stress and stress amplitude, and conventionally presented their results as constant life diagrams showing the relationship between mean stress and stress amplitude (Fig. 8). In every case the criterion of failure was separation of the specimens, but two modes of failure were noted, tensile and compressive. Tensile failures were somewhat hairy failures normal to the axis of the specimens whereas compressive failures were smooth failures inclined at approximately 45 deg to the axis of the specimen. For those combinations of mean stress and stress amplitude where short life failures were compressive and long life failures were tensile, it was found that there was a marked knee or discontinuity in the S-N curve (Fig. 9), whereas if only one mode of failure occurred the S-N curve was essentially a straight line. The most important conclusion from the master diagrams was that the Goodman law, commonly used for structural metals, was unconservative for CSM/PR (Fig. 8). Owen and Smith [10] made a comparison between fatigue and stress rupture data

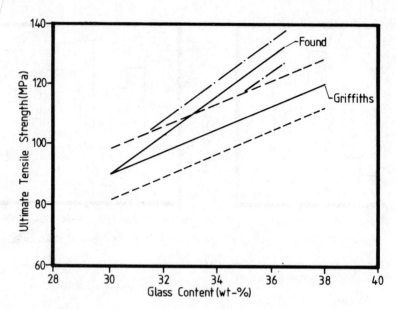

FIG. 4—*Relationship between glass content and UTS* [9]. *Found: 30 results; correlation coefficient,* +0.80; *standard error of estimate,* ±4.24 MPa. *Griffiths: 43 results; correlation coefficient,* +0.80; *standard error of estimate,* ±6.54 MPa.

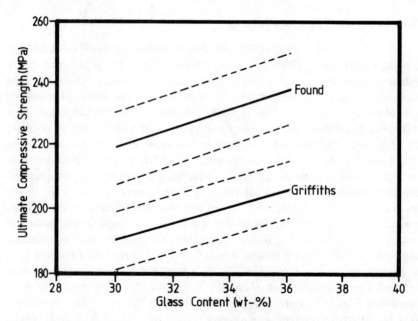

FIG. 5—*Relationship between glass content and UCS* [9]. *Found: 18 results: correlation coefficient,* +0.27; *standard error of estimate,* ±10.63 MPa. *Griffiths: 33 results; correlation coefficient,* +0.32; *standard error of estimate,* ±10.06 MPa.

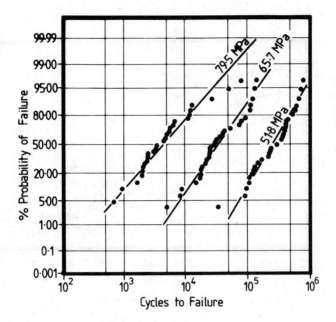

FIG. 6—*Log-normal distribution of fatigue lives for CSM/PR* [3].

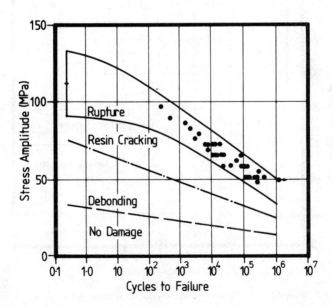

FIG. 7—*Fatigue data for CSM/PR specimens at zero mean stress* [10]; *stress by the UTS of the nearest neighbor in the laminate* [5,6].

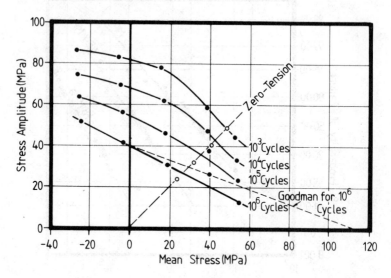

FIG. 8—*Master diagram for CSM/PR specimens* [10].

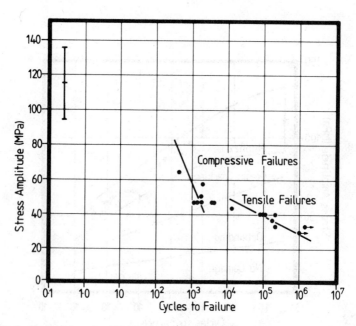

FIG. 9—*Fatigue data for CSM/PR specimens at 42.0 MPa mean stress showing effect of failure mode* [10].

and concluded that repeated applications of load (fatigue) were far more damaging than the same load applied for a sustained period of time (Fig. 10). Subsequently Owen and Howe [6] showed that there was an interaction between fatigue damage and stress rupture damage. By carrying out fatigue loading to various fractions of the expected mean fatigue life and then carrying out stress rupture tests they showed that the prior fatiguing produced a marked reduction in the subsequent stress rupture life (Fig. 11). It is this effect that undoubtedly makes the Goodman law unconservative in expressing the mean stress-stress amplitude relationship for CSM/PR.

Owen et al [3] produced S-N curves which showed the onset of transverse fiber debonding and the onset of resin cracking (Fig. 12). They also kept track of reductions in the initial modulus of the specimens due to damage. They further showed that the onset of debonding under a single application of tensile load occurred at approximately 0.3 percent strain for a wide variety of GRPs, and furthermore that at the onset of debonding, S-N curves for widely differing GRPs were almost superimposed on a strain basis.

A major omission in some early published work was the failure to include elastic constants with the fatigue results. This failure stems from adhering to the practices adopted for metals. Very few fatigue papers on metals include

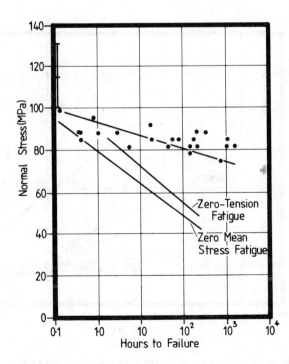

FIG. 10—*Stress rupture data compared with fatigue curves* [10].

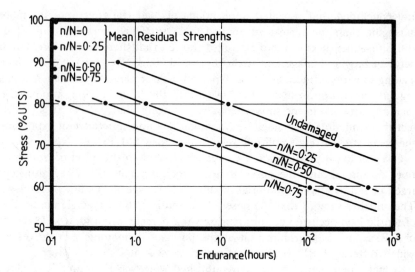

FIG. 11—*Effect of prior fatigue damage on subsequent stress-rupture performance. Cycle ratio* n/N *is based on mean expected fatigue life* [6].

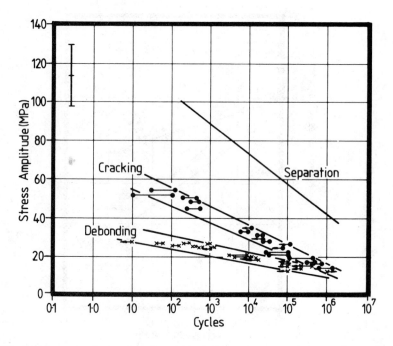

FIG. 12—*The onset of transverse fiber debonding and resin cracking in fatigue under zero mean stress conditions* [3].

elastic constants. The implication is that stress analysis can be carried out using conventional values for ferrous and aluminum alloys. This is not the case
for GRP. It is highly desirable that any stock of material used for fatigue
testing should be fully characterized with its static strength and the initial
values of the elastic constants.

Effect of a Stress Concentrator

Owen and Bishop [11,12] studied the effect of a stress concentrator on a
wide variety of FRPs, including CSM/PR. Their object was to correlate the
static and fatigue properties of plain specimens with the corresponding properties of specimens containing a round hole. The correlation was effected by
means of finite-element stress analysis for finite width of plates sup-
plemented by Savin's equations [13] for plates of infinite width. In the special
case of CSM/PR the material was treated as plane isotropic. In all the property determinations they observed the onset of debonding and the onset of
resin cracking as well as the final rupture of the specimens. Figure 13 shows
static strength scatter bands and S-N curves for the three states of failure for
plain specimens and specimens with holes. Table 1 summarizes some of the
results obtained by Owen and Bishop [11,12]. In essence they found that a

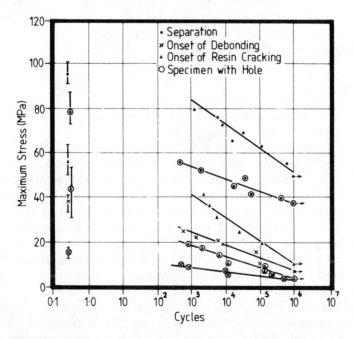

FIG. 13—*Zero-tension stress fatigue results for CSM/PR specimens both with and without a
hole* [12].

TABLE 1—*Effect of a hole in a 2.54-cm-wide parallel strip of CSM/PR* [11,12].[a]

Parameter	Plain Specimen(a)	Specimen with Hole(b)	a/b
Static UTS	137.0	78.2	1.75
Static debonding	33.4	15.4	2.17
10^6 cycles rupture	52.0	36.4	1.43
10^6 cycles debonding	6.90	3.44	2.01

[a]Elastic stress concentration factor of 3.2. Observed strengths are given in MPa.

small hole in CSM/PR was not a fully effective stress concentrator. However, in checking the stress distribution with strain gages with a specimen ten times larger they found that the strength of the large specimen was only half that of the smaller one. They also noted that the onset of debonding at the edge of a hole of 10^6 cycles occurs at a very small fraction of the nominal stress representing ultimate failure of the plain material, and concluded that for most purposes the onset of debonding at a hole was uneconomic as a failure criterion. They noted two other important factors. The *S-N* curve between 10^3 and 10^6 cycles can usually only be represented as a straight line that extrapolates to zero stress at a finite life, thus making it very difficult to use the *S-N* curves for predicting safe loads for long life times. These observations together with the size effect noted above prompted them to look at fracture mechanics applied to CSM/PR.

Biaxial Strengths

Real engineering components and structures are hardly ever subjected to simple uniaxial loading. For many FRPs, the loading conditions produce plane stress. Owen and Found [14], Owen and Griffiths [15], and Owen et al [9] have carried out numerous biaxial stress tests on thin-walled tubes subjected to combined internal pressure and axial loading. The object of this work was to evaluate failure theories for multiaxial stress. Over 40 theories of failure (functions of stresses and strengths) have been proposed and applied to FRP. At first sight CSM/PR are plane isotropic, but in general their tensile and compressive strengths are not equal and so at least in this respect they are anisotropic. Figure 14 shows a plot of hoop stress against axial stress at debonding, resin cracking, and separation and bursting for thin-walled tubes at various principal stress ratios (R). The results at $R = +\infty$ and $R = -\infty$ are obtained from conventional flat specimens. Figure 15 shows constant life diagrams for similar fatigue tests, and Fig. 16 compares static and fatigue tests at the onset of resin cracking. In the tension-tension quadrant the results will fall well inside a maximum principal stress or maximum prin-

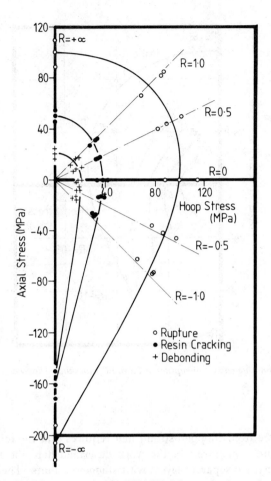

FIG. 14—*Static biaxial stress results for CSM/PR tubes under internal pressure and axial load* [14].

cipal strain theory, and furthermore the tension-tension condition is particularly damaging in fatigue. When the hoop stress is greater than the axial stress, resin cracking is always parallel to the tube axis (that is, like the damage in a simple tension specimen). When the hoop and axial stresses are equal, however, the resin cracking damage forms a network [9]. It is this form of damage that appears to be so damaging in fatigue. Among the major difficulties in fitting failure theories to the results is the fact that some failure theories require in-plane shear strengths both under static and fatigue loading. Such results are extremely difficult to obtain with any degree of confidence. They are markedly affected by the specimen geometry. Another major difficulty results from the presence of joints in the specimen. To make a

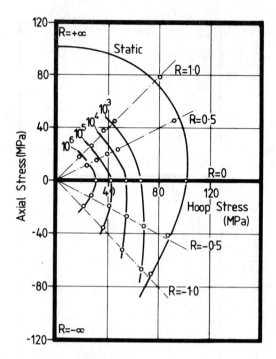

FIG. 15—*Constant life curves for rupture of CSM/PR tubes under internal pressure and axial load* [14].

thin-walled tube from chopped strand mat requires that the reinforcement is wrapped around a mandrel. In the work described here the reinforcement was applied in three separate layers with staggered joints. These joints were virtually obliterated during the laminating procedure, and the static hoop strength of tubes coincided with the static tensile strength of flat specimens. However, Fig. 17 shows *S-N* curves for flat specimens and tubes at $R = 0$ (that is, pure hoop stress conditions); it is seen that the fatigue behavior of the tubes is very much inferior to the fatigue behavior of the flat specimens [9]. The figure also reveals that there were marked differences between the behavior of two different batches of reinforcement to the same specification. By carrying out fatigue tests on flat specimens containing similar joint geometries it was found that the discrepancy was due to the presence of the joints in the layers of reinforcement [9]. By observation it was found that under fatigue loading resin cracking damage occurred prematurely at the joint in the mat layer and that this propagated through the adjacent continuous layers to promote early failure. Large structures made from CSM/PR inevitably contain such joints because they are made from rolls of finite width.

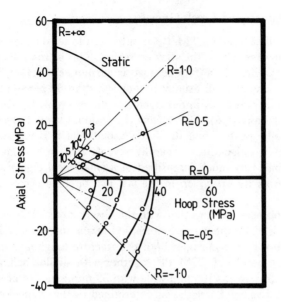

FIG. 16—*Constant life curves for resin cracking in CSM/PR tubes under internal pressure and axial load* [14].

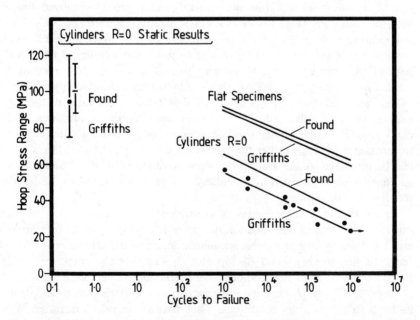

FIG. 17—*Fatigue results for thin-walled tubes subject to hoop stress only compared with fatigue results for flat specimens of CSM/PR. As in Figs. 6 and 7, there are differences between two matches of equivalent material identified by the names of two experimenters* [9].

Fracture Toughness and Crack Growth

When a small sample of CSM/PR is subjected to tensile load, damage appears uniformly throughout the stressed region. This is true under both static and fatigue loading. If a fatigue test is terminated before fracture the residual static strength will usually be not less than 85 percent of the mean static strength of previously undamaged specimens even when the cycle ratio is approaching unity [5,6]. This means the residual static strength is generally substantially greater than the fatigue load that is being applied. Close observation of a fatigue test generally reveals that in the last few cycles damage becomes very much intensified in a small local region and that this propagates across the specimen in the last two or three cycles almost like a blunt crack.

The above observation, together with the observations of Owen and Bishop on the behavior of specimens with holes, led Owen and Bishop [16] and later Owen and Cann [17] to consider both the fracture toughness and the crack propagation properties of CSM/PR together with similar behavior of other types of FRP. Specimens 100 mm wide with 33 mm long center notches were used. The effective crack length was determined by the change of compliance using previously calibrated specimens. Crack growth experiments were carried out under constant crack tip stress intensity factor range (ΔK) loading. Initially the crack growth appeared to change rapidly while a damage zone formed at the crack tip, but subsequently crack growth occurred linearly with increasing life (Fig. 18). Stable crack growth occurred over only a narrow range of ΔK values. At the smaller values crack arrest tended to occur from time to time; this could be related to particular bundles of fiber in the path of the growing crack. Owen and Bishop [16] showed by plotting log crack growth rate against log ΔK that the behavior conformed approximately to the Paris [18] crack growth law $dA/dN = C\Delta K^m$. The value of the exponent m was unusually high, 12.75; hence the difficulty in obtaining stable crack growth. Under service conditions materials are more usually subjected to constant load range rather than constant ΔK range, but the effects of this can be predicted by integrating the Paris equation (Fig. 19). The life of the specimen is then governed by unstable crack growth rather than the fracture toughness of the material.

Figure 19 shows three values of equivalent crack growth. Any of these might be used as a design criterion that would then allow for the formation of a stable damage zone at a stress concentrator and would also allow for size effects. In Ref 16 Owen and Bishop also showed that the crack growth approach could be used to extrapolate life predictions to very long lives.

Owen and Cann [17] carried out crack growth experiments to longer lives under both wet and dry conditions. They found that prior saturation of the specimens in distilled water produced an increase of approximately two orders of magnitude in crack growth rate. They also examined the ap-

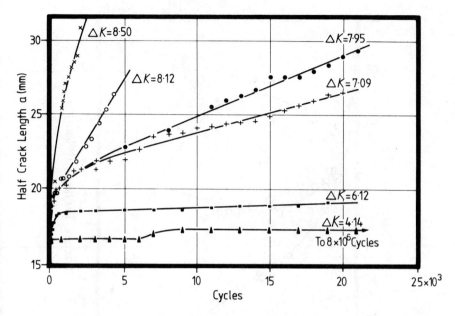

FIG. 18—*Crack growth in CSM/PR for various values of* ΔK *(MN m$^{-3/2}$)* [16].

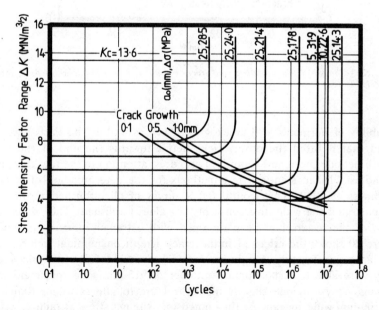

FIG. 19—*Failure under constant load range for CSM/PR* ($K_c = 13.6$ *MN m$^{-3/2}$) for 100-mm-wide specimens with various initial crack lengths* a_0 *and initial stress ranges* $\Delta\sigma$.

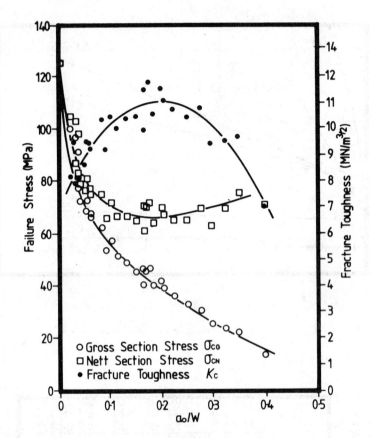

FIG. 20—*Change of fracture stress and fracture toughness with notch width for CSM/PR material.*

plicability of Forman's [*19*] crack growth law as well as the Paris law, and found that it fitted the observations better towards the end of the crack growth period. Owen and Cann [*17*] also drew attention to the fact that there are no established standard test methods for the determination of fracture toughness of GRP. They conducted a survey of all published results that demonstrated the wide divergence of the values published. They noted that many of the results had been obtained with extremely narrow specimens. Figure 20 shows the effect of initial crack length on nominal failure stress and apparent fracture toughness for 100-mm-wide CSM/PR specimens. They were able to test a limited number of 915-mm-wide specimens and, although they obtained slightly higher fracture toughness values than with the 100-mm-wide specimens, they noted that the net stress at failure was extremely low compared with the normally accepted values of UTS.

Conclusion

Glass chopped strand mat/polyester resin combinations are a useful engineering material, although there are a number of pitfalls in using it for engineering purposes. There is considerable scatter in its properties. In the laboratory this scatter can be controlled by controlling thickness and glass content, but this is more difficult to do in the industrial shop. Although CSM/PR is nominally plane isotropic, the failure behavior is governed by the anisotropic nature of the strand. There are marked adverse size effects on strength that are not particularly well documented. The fracture toughness is generally low, and once damage has been initiated large scale cracks can propagate readily. Reinforcement joints may well promote cracking. Further work on fracture mechanics and crack growth analysis would be highly desirable.

References

[1] Owen, M. J. and Dukes, R., *Journal of Strain Analysis*, Vol. 2, No. 4, Oct. 1967, pp. 272-279.
[2] Owen, M. J., Dukes, R., and Smith, T. R., "Fatigue and Failure Mechanisms in GRP with Special Reference to Random Reinforcements," in *Proceedings*, 23rd Annual Reinforced Plastics Conference, Society of the Plastics Industry, Washington, D.C., Feb. 1968, Section 14-A.
[3] Owen, M. J., Smith, T. R., and Dukes, R., *Plastics and Polymers*, Vol. 37, June 1969, pp. 227-233.
[4] Owen, M. J. and Rose, R. G., "The Effect of Resin Flexibility on the Fatigue Behavior of GRP," Paper 8, British Plastics Federation 7th International Reinforced Plastics Conference, Brighton, October 1970; also published in *Modern Plastics*, Vol. 47, No. 11, Nov. 1970, p. 130.
[5] Owen, M. J. and Howe, R. J., *Journal of Physics (D): Applied Physics*, Vol. 5, 1972, pp. 1637-1649.
[6] Howe, R. J. and Owen, M. J., "Cumulative Damage in Chopped Strand Mat/Polyester Resin Laminates," 8th International R. P. Conference, British Plastics Federation, Brighton, Oct. 1972, Paper 21.
[7] Owen, M. J. and Rose, R. G., *Journal of Physics (D): Applied Physics*, Vol. 6, 1973, pp. 42-53.
[8] Howe, R. J., "Cumulative Damage of a Glass Reinforced Plastic," Ph.D. thesis, University of Nottingham, U.K., 1971.
[9] Owen, M. J., Griffiths, J. R., and Found, M. S., "Biaxial Stress Fatigue Testing of Thin-Walled GRP Cylinders," in *Proceedings*, 1975 International Conference on Composite Materials, Vol. 2, American Institute of Mining, Metallurgical, and Petroleum Engineers, New York, 1976, pp. 917-941.
[10] Owen, M. J. and Smith, T. R., *Plastics and Polymers*, Vol. 36, Feb. 1968, pp. 33-44.
[11] Owen, M. J. and Bishop, P. T., *Journal of Physics (D): Applied Physics*, Vol. 5, 1972, pp. 1621-1636.
[12] Owen, M. J. and Bishop, P. T., *Jounal of Physics (D): Applied Physics*, Vol. 6, 1973, pp. 2057-2069.
[13] Savin, G. N., *Stress Concentration around Holes*, Pergamon Press, London, 1961.
[14] Owen, M. J. and Found, M. S., "Static and Fatigue Failure of Glass Fibre Reinforced Polyester Resins Under Complex Stress Conditions," in *Faraday Special Discussions of the Chemical Society*, No. 2, 1972, pp. 77-89.
[15] Owen, M. J. and Griffiths, J. R., *Composites*, April 1979, pp. 89-94.

[16] Owen, M. J. and Bishop, P. T., *Journal of Physics (D): Applied Physics,* Vol. 7, 1974, pp. 1214-1224.
[17] Owen, M. J. and Cann, R. J., *Journal of Material Science,* Vol. 14, 1979, pp. 1982-1996.
[18] Paris, P. C. and Sih, G. C., in *Fracture Toughness Testing and Its Applications, ASTM STP 381,* American Society for Testing and Materials, 1965, pp. 30-83.
[19] Forman, R. G., Kearny, V. E., and Engle, P. M., *Journal of Basic Engineering,* Vol. 89, 1967, pp. 459-464.

E. M. Caulfield[1]

An Investigation of Stress-Dependent, Temperature-Dependent, and Time-Dependent Strains in Randomly Oriented Fiber Reinforced Composites

REFERENCE: Caulfield, E. M., **"An Investigation of Stress-Dependent, Temperature-Dependent, and Time-Dependent Strains in Randomly Oriented Fiber Reinforced Composites,"** *Short Fiber Reinforced Composite Materials, ASTM STP 772,* B. A. Sanders, Ed., American Society for Testing and Materials, 1982, pp. 85–96.

ABSTRACT: The effect on modulus change due to temperature on the stress-strain response of a glass fiber reinforced composite is discussed. Operationally simple constitutive equations for stress and strain computation while material temperature is transient are presented; these equations employ stress, internal specimen time, coefficient of thermal expansion, and measured material constants to determine the stress-strain response of fiberglass composites. Formulation of generalized constitutive equations to particular stress temperature histories is adduced and experimentally verified.

KEY WORDS: total strain, thermal strain, constitutive, internal time, temperature-shifted real time, internal variable, viscoelastic strain, creep compliance, relaxation modulus, apparent modulus, slope of stress-strain curve, coefficient of free thermal expansion

In recent years, challenging design situations have evolved from the fabrication of containment vessels for use at cryogenic temperatures. Since the materials used in the construction of these vessels may be subjected to temperature changes of approximately 225 K, the effects of changes in Young's modulus and the coefficient of thermal expansion changing with temperature must be considered in determining the stresses and strains present in the structural members.

---

[1]Director of Mechanical Engineering, Packer Engineering Associates, Naperville, Ill. 60540.

The temperature dependence of Young's modulus and the coefficient of thermal expansion is readily observed in glass fiber reinforced plastic (GFRP), a material of considerable interest for cryogenic containment vessels. It is empirically known that some GFRPs exhibit a linear stress-strain relation from room temperature (300 K) to liquid nitrogen temperature (78 K). It is also known that Young's modulus may change by a factor of two over this temperature range. Similarly, the coefficient of thermal expansion, which relates thermal strain to temperature change, continually decreases from 300 to 78 K. Even though the relative change in the coefficient of thermal expansion from 300 to 78 K is small in magnitude compared with the relative change in Young's modulus over the same temperature range, the approximation that the coefficient of thermal expansion is not a function of temperature is an oversimplification.

As a result of variations with temperature of the properties mentioned above, accurate predictions of stress and strain under varying temperature conditions become complicated. Much of the past work on the mechanical behavior of structural materials subjected to temperature changes has been in the area of thermal stress. The conventional approach to the problem, however, assumes the coefficient of thermal expansion and Young's modulus to be constant over the temperature range considered. Johns [1], Hilton, [2], Chang and Chu [3], Newman and Forray [4], and Baltrukonis [5] considered Young's modulus as a function of temperature for materials with through-thickness temperature gradients.[2] These authors, however, considered temperature-increasing situations and, apparently, none of their work has been experimentally verified.

This paper presents operationally simple constitutive equations that predict the mechanical response of GFRP subjected to varying thermal conditions.

Procedure

In order to effectively study the mechanical response of a glass fiber reinforced composite, a closed-loop hydraulic test machine was employed. An important feature of the closed-loop test system lies in its ability to control load, strain, or cross-head displacement by continually monitoring and correcting these parameters with high-speed electronic circuitry. In addition, the control of the rate of application of these parameters to a specimen may be accomplished by the addition of an electronic signal generator to the closed-loop system. The electronic control mechanism together with the hydraulic actuator enables the test system to exhibit high-speed parameter control. Figure 1 is a photograph of the test system employed.

---

[2]The italic numbers in brackets refer to the list of references appended to this paper.
[3]Original experimental data were given in English customary units.

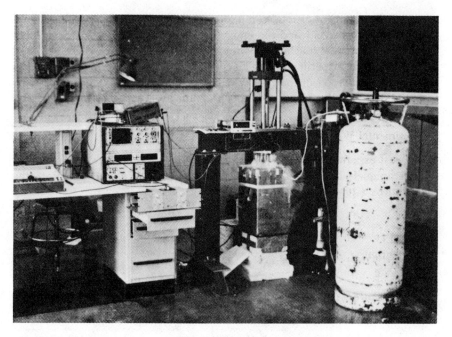

FIG. 1—*Test system.*

Thermal Chamber

The temperature range under consideration was 300 to 78 K. An insulated test chamber in conjunction with a thermally controlled liquid nitrogen flow system was used to obtain a controlled temperature environment. By employing this apparatus, any thermal environment in the range of 300 to 78 K could be controlled within ±2 K.

Strain Extensometer

An extensometer constructed of fused silica quartz was developed for strain measurement. The choice of fused silica quartz resulted from its coefficient of thermal expansion being approximately 2.2×10^{-7} in./in./K, which is a factor of 100 less than that of the glass fiber reinforced composite being evaluated. As a result, all thermal strains due to the expansion/contraction of the fused silica extensometer were neglected in all measurements throughout the experiment.

Specimens and Materials

The material employed throughout the study was glass fiber reinforced plastic consisting of randomly oriented chopped glass fibers in an epoxy matrix. On a percent of total composite volume basis, the glass made up approximately 30 percent and the epoxy approximately 70 percent. Specimens of this material were machined from a 6.35-mm (¼-in.)-thick sheet and used throughout the experiment. No particular attention was directed towards specimen orientation in that the orientation was randomly selected throughout the sheet. The specimen geometry and dimensions are illustrated in Fig. 2.

Experimental Work

In past experimental work [6], it has been shown that the strain in GFRP is not uniquely dependent on stress and temperature alone. Therefore the constitutive equation for strain must be written as

$$\epsilon = \ell\,(\sigma, T, C)$$

Differentiating, one obtains

$$d\epsilon = \left(\frac{\partial \epsilon}{\partial \sigma}\right)_{T,C} d\sigma + \left(\frac{\partial \epsilon}{\partial T}\right)_{\sigma,C} dT + \left(\frac{\partial \epsilon}{\partial C}\right)_{\sigma,T} dC \qquad (1)$$

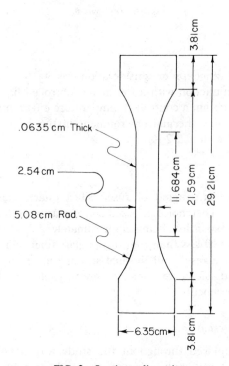

FIG. 2—*Specimen dimensions.*

In order to experimentally evaluate the parameters in Eq 1, it is necessary to employ a test system that is capable of controlling stress σ, temperature T, and the internal variable C. The test system used throughout the study has no difficulty controlling stress and temperature, but C must first be determined before an effort is made toward controlling this parameter.

Through experimentation, it has been determined that C corresponds to the temperature-shifted real time ξ defined as

$$d\xi = \frac{dt}{[a(T)]} \tag{2}$$

and

$$\xi = \int_0^t \frac{dt}{[a(T)]} \tag{2a}$$

A general result for differential strain compatible with Eq 1 has been determined to be [7]

$$d\epsilon = \frac{1}{E_0} d\sigma + \alpha dT + \left[\frac{\sigma_c}{E_1} \xi^{N-1} N \right]_{\sigma,T} d\xi \tag{3}$$

where

E_0 = a constant,
α = coefficient of free thermal expansion,
σ_c = applied nominal stress,
t = time,[4]
ξ = temperature-shifted real time or internal time,[4]
N = a constant, and
E_1 = a constant.

Assuming the Boltzmann superposition principle [8], a general equation for total strain employing Eq 3 may be written as

$$\epsilon = \int_0^\sigma \frac{d\sigma}{E_0} + \int_0^\xi \frac{1}{E_1} \left(\xi - \xi' \right)^N \frac{d\sigma}{d\xi'} d\xi' + \int_{T_1}^T \alpha dT \tag{4}$$

where

ξ' = a dummy variable of integration corresponding to ξ,
ξ = an arbitrary internal time,
T_1 = initial temperature, and
T = temperature at internal time ξ.

---

[4]Time t is measured in a time frame such that $t = 0$ upon load application. This definition of t will be used throughout. Moreover, it will be assumed throughout that all time parameters will be dimensionless and measured in minutes.

It can be shown that Eq 4 is compatible with Eq 3 and it will therefore be used throughout the remainder of the paper as the constitutive equation for the material under consideration.

Experimental Evaluation of Parameters

If the applied stress is a linear function of time and the test temperature is constant, that is,

$$\sigma = Kt = K\xi[a(T)] \qquad (T = \text{constant})$$

where K is a proportionality constant, then Eq 4 may be integrated to obtain

$$\epsilon = \frac{\sigma}{E_0} + \frac{\sigma}{E_1} \xi^N \frac{1}{N+1} \qquad (T \text{ constant}: \sigma = Kt) \qquad (5)$$

The first parameter that warrants empirical evaluation is the constant E_0. Due to the magnitude of the temperature shift-factor at low temperature, the value of ξ becomes very small in Eq 5. The low-temperature condition employed to evaluate E_0 was 78 K. Numerically, E_0 was found to be 18.7 GPa (2.72×10^6 psi). It is worthwhile noting that the value of σ/ϵ from 90 to 78 K is essentially constant, which implies that at these temperatures the internal time ξ is becoming vanishingly small. Moreover, extrapolation from 78 to 0 K for the value of σ/ϵ is unrevealing because of the small changes in σ/ϵ from 90 to 78 K.

The parameters E_1 and N may be evaluated at room temperature. Without loss in generality, the value of the shift-factor $[a(T)]$ may be set equal to unity and correspondingly ξ may be replaced by t in the appropriate constitutive equations.

Since step functions of stress are impossible to impose in the laboratory, the stress applied to the material will initially be a linear function of time until the stress reaches a desired value; it will then be held constant for the remainder of the test. If the rise time for stress is small compared with the time duration at constant stress, then Eq 4 will reduce to

$$\epsilon = \frac{\sigma_c}{E_0} + \frac{\sigma_c}{E_1} t^N \qquad (T \text{ constant}) \qquad (6)$$

where σ_c is the constant value of stress being investigated. The immediate advantage of making the stress rise time small results from the operationally simple nature of Eq 6.

Accordingly, constant stress creep tests with rapid stress rise times were performed and the values of E_1 and N were found to be

$$E_1 = 26.3 \text{ GPa } (3.82 \times 10^6 \text{ psi})$$

$$N = 0.07$$

It is noted that the time-dependent strains in this experiment are linearly dependent on stress, which is in agreement with the Boltzmann superposition principle assumed earlier. With the parameters E_0, E_1, and N evaluated, the time-temperature shift-factor was investigated. Equation 5 may be solved for the quantity ξ if one knows ϵ, σ, E_0, E_1, and N. The internal time ξ may be described using an Arrhenius-type equation

$$\xi = \frac{t}{[a(T)]} = t\,e^{\,Q(1/T_0 - 1/T)}$$

where Q is a material constant. Since T_0 is room temperature (300 K) and t and T are known for stress rate tests over a variety of temperatures, this equation may be solved for Q using stress-strain curves obtained at different temperatures. Due to the sensitive behavior of the Arrhenius exponential relation above, the value of Q was found to be 5000 ± 500 K^{-1}. This deviation of ± 500 K^{-1} may be tolerated in the experimental results because of the small value of the exponent N on the ξ-parameter in the constitutive equations.

Results

Strain Resulting from Temperature Change with Stress Being Held Constant

This stress-strain response employs Eq 4 because the temperature is variable. A stress-temperature sequence to employ Eq 4 is outlined as follows: (*1*) decrease the temperature from 300 to 78 K with $\sigma = 0$; (*2*) stress the material to a predetermined stress σ_c; (*3*) increase the temperature back to 300 K holding the stress constant at σ_c; and (*4*) decrease the temperature to 78 K while stress is held constant at σ_c. Step 1 of this sequence determines the free coefficient of thermal expansion and needs no further discussion. The resulting strain after Step 2 is

$$\epsilon_2 = \frac{\sigma_c}{E_0} + \frac{\sigma_c}{E_1}\,\xi^N \left(\frac{1}{N+1}\right) + \int_{300}^{78} \alpha\,dT$$

Due to the large value of $[a(T)]$ at 78 K, ξ becomes vanishingly small and the resulting strain may be approximated by

$$\epsilon_2 \approx \frac{\sigma_c}{E_0} + \int_{300}^{78} \alpha\,dT$$

After Step 3 is completed, the resulting strain may be shown to be

$$\epsilon_3 = \frac{\sigma_c}{E_0} + \frac{\sigma_c}{E_1}\,\xi^N$$

The value of ξ from Step 2 to Step 3 may be approximated by the time spent at room temperature, because of the magnitude of the shift-factor $[a(T)]$ across the temperature excursion. Therefore the strain after Step 3 may be written as

$$\epsilon_3 \simeq \frac{\sigma_c}{E_1} + \frac{\sigma_c}{E_1} t_R{}^N$$

where t_R is the time spent at room temperature. Step 4 is the reverse of Step 3 in procedure and the resulting calculation for strain may be shown to be

$$\epsilon_4 \simeq \frac{\sigma_c}{E_0} + \int_{300}^{78} \alpha dT + \frac{\sigma_c}{E_1} \left(t_R + \xi \right)^N$$

Again, due to the magnitude of the shift-factor in the parameter ξ, it can be shown that $\xi \ll t_R$; hence the resulting equation for ϵ_4 is

$$\epsilon_4 \simeq \frac{\sigma_c}{E_0} + \int_{300}^{78} \alpha dT + \frac{\sigma_c}{E_1} t_R{}^N$$

Comparing ϵ_4 with ϵ_3 and ϵ_2, we note that (1) the strain at Steps 3 and 4 only differs by the amount

$$\int_{300}^{78} \alpha dT$$

and (2) the strain at Steps 4 and 2 differs by an amount approximately equal to

$$\frac{\sigma_c}{E_1} t_R{}^N$$

Recalling that the material at Steps 4 and 2 is at the same stress and temperature values, the difference in total strain is attributed to the time-dependent strain being irreversible; that is, $d\xi$ and dt are always greater than zero.

The experimental and predicted strain responses for the total sequence, described by Steps 1 to 4, are shown in Fig. 3.

Strain Resulting from a Controlled Stress that is Linearly Dependent on Time

A fundamental stress-strain behavior results from controlling stress as a function of time and measuring strain; moreover, plotting stress versus strain results in the "stress-strain curve", which is an important tool in material analysis.

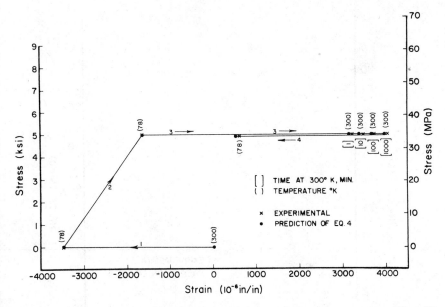

FIG. 3—*Stress-strain response resulting from* (1) *decreasing the temperature from 300 to 78 K,* (2) *stressing to 34.45 MPa (5 ksi),* (3) *increasing the temperature from 78 to 399 K, and* (4) *decreasing the temperature from 300 to 78 K (1 in. = 25.4 mm).*

In order to predict the strain response for this stress control condition, Eq 5 is employed. For a temperature of 200 K, the shift-factor [$a(t)$] becomes

$$e^{5000[1/200 - 1/300]} = 4160$$

and Eq 5 becomes

$$\epsilon = \frac{\sigma}{E_0} + \frac{\sigma}{E_1}\left(\frac{t}{4160}\right)^N\left(\frac{1}{N+1}\right)$$

The stress-strain responses at 300, 200, and 78 K are shown with predictions from Eq 5 in Figs. 4, 5, and 6, respectively. Again, the empirical stress-strain result may be represented by a straight line, but this line exhibits a higher slope as temperature decreases.

Summary and Conclusions

The predicted and experimental results exhibit close agreement throughout the stress-temperature-strain histories examined. The small deviations between these results may easily be attributed to simplifying assumptions made in the predictions and to experimental scatter. Nonetheless, as a result

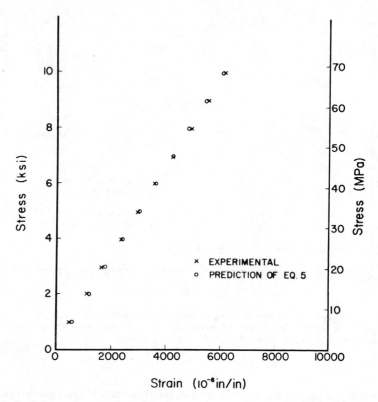

FIG. 4—*Stress-strain response resulting from a stress rate of 68.9 MPa/min (10⁴ psi/min) at 300 K (1 in. = 25.4 mm).*

of the agreement between predicted and experimental results, the glass fiber reinforced composite used in this study is best analyzed in the temperature region of 300 to 78 K by time-dependent constitutive equations.

Employing the methods developed in this paper, the author has predicted the strain response of glass fiber reinforced composites over wide temperature ranges. The operationally simple constitutive equations employ stress and temperature as functions of time to predict strain. The strain response is determined for the entire stress-temperature history and does not sum components of strain for individual stress-temperature events. It is seen from the results that the linearity of the stress-strain response of GFRP is predictable using time-dependent constitutive equations. Accordingly, the assumption that GFRP is an elastic material due to the stress-strain linearity is an erroneous oversimplification for widely varying thermal conditions.

The operationally simple constitutive equations presented here answer many perplexing questions about the complicated behavior of glass fiber

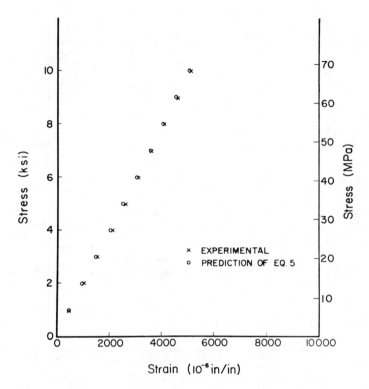

FIG. 5—*Stress-strain response at 200 K resulting from a stress rate of 68.9 MPa/min (10⁴ psi/min) (1 in. = 25.4 mm).*

reinforced composites subjected to stress-temperature histories. Recent problems in cryogenic design situations have generated the need to accurately describe the mechanical response of these materials over wide temperature ranges. The experience and data obtained from cryogenic applications of GFRP design situations have been and will continue to be a valuable aid in verifying and refining the constitutive equations presented here.

Acknowledgments

I wish to thank University of Illinois professors E. C. Aifantis, G. A. Costello, R. E. Miller, J. Morrow, G. M. Sinclair, and the late M. C. Stippes for valuable technical discussions throughout the development of this research. I also wish to acknowledge Owens Corning Fiberglas Corporation of Granville, Ohio, and the Fermi Laboratory of Batavia, Illinois, which have contributed to the general development of this project.

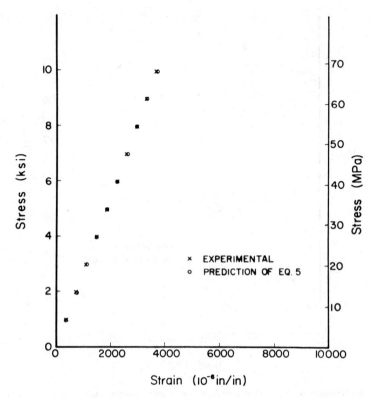

FIG. 6—*Stress-strain response at 78 K resulting from a stress rate of 68.9 MPa/min (10^4 psi/min) (1 in. = 25.4 mm).*

References

[1] Johns, D. J., *Thermal Stress Analysis*, Pergamon Press, Oxford, 1965.
[2] Hilton, H. H., *Journal of Applied Mechanics*, Vol. 19, 1952, pp. 350-354.
[3] Chang, C. C. and Chu, W. H., *Journal of Applied Mechanics*, Vol. 21, 1954, pp. 101-108.
[4] Newman, M. and Forray, M., *Journal of the Aero/Space Sciences*, Vol. 28, 1962, pp. 372-373.
[5] Baltrukonis, J. H., *Journal of the Aero/Space Sciences*, Vol. 26, 1959, pp. 329-334.
[6] Caulfield, E. M., "An Investigation of Stress, Temperature, and Time Dependent Strains in a Randomly Oriented Fiber Reinforced Composite with Special Emphasis Given to Thermal Stress Situations," T&AM Report 432, Department of Theoretical and Applied Mechanics, University of Illinois, Urbana, 1979, pp. 9-14.
[7] Caulfield, E. M., "An Investigation of Stress, Temperature, and Time Dependent Strains in a Randomly Oriented Fiber Reinforced Composite with Special Emphasis Given to Thermal Stress Situations, T&AM Report 432, Department of Theoretical and Applied Mechanics, University of Illinois, Urbana, 1979, pp. 17-18.
[8] Broutman, L. J. and Krock, R. H., *Modern Composite Materials*, Addison-Wesley, Reading, Mass., 1967, p. 125.

J. C. Duke, Jr.[1]

Nondestructive Characterization of Chopped Glass Fiber Reinforced Composite Materials

REFERENCE: Duke, J. C., Jr., **"Nondestructive Characterization of Chopped Glass Fiber Reinforced Composite Materials,"** *Short Fiber Reinforced Composite Materials, ASTM STP 772,* B. A. Sanders, Ed., American Society for Testing and Materials, 1982, pp. 97–112.

ABSTRACT: Quality control methods are insufficient for assuring defect-free composite materials, which are manufactured at the time of part fabrication. Nondestructive evaluation (NDE) techniques appear, potentially, to offer a means of identifying such defects. The techniques of ultrasonic C-scanning, X-ray radiography, vibrothermography, and acoustic emission monitoring are considered here. The materials of interest are E-glass and S-2 glass reinforced sheet molding compound. Results from ultrasonic C-scans clearly indicate areas of imperfections. Results of the vibrothermography and acoustic emission monitoring, however, suggest the need for additional work in these areas, while those of X-ray radiography are inconclusive.

KEY WORDS: nondestructive evaluation, ultrasonic C-scan, X-ray radiography, vibrothermography, acoustic emission, composite materials, sheet molding compound

Combination of materials into composite systems has made possible the solution of materials problems previously unsolved, as well as providing better solutions to problems already answered. Of particular benefit to those areas concerned with energy savings are the high specific strength and specific stiffness characteristics of some composite systems. However, due to the inherent nature of composite materials, in general their properties are extremely sensitive to the materials' microstructure. Variations in distribution, orientation, or amount of constituents from point to point within the material can give rise to large differences in material properties. The problem is further compounded by the fact that the composite material used in the fabrication of parts or structures is in many cases manufactured at the

[1]Assistant Professor, Materials Response Group, Engineering Science and Mechanics Dept., Virginia Polytechnic Institute and State University, Blacksburg, Va. 24061.

time of the fabrication. Any material defects are thus incorporated in the product, allowing no chance for elimination prior to fabrication. At present, methods of quality control processing are insufficient to assure the integrity of many products. It is necessary, therefore, that additional steps be taken so that design-specified performance is not compromised by material defects.

Nondestructive evaluation (NDE) techniques appear to offer a potential means of identifying such defects. However, efforts to develop such techniques can ultimately be of use only if the results can be related to the material's performance; that is, an imperfection, regardless of shape, size, orientation, etc., that does not affect the material's response in service is of no consequence.

The present study has made an effort to evaluate several different NDE techniques and relate the findings to the material's performance for chopped glass fiber reinforced polyester systems. Special consideration was given to feasibility as well as ease of applicability for industrial use when selecting the techniques to be evaluated in this study.

Description of Materials

The materials tested in this study were made by the Owens-Corning Fiberglas Corporation. They included: 1-in. E-glass R65 SMC, 2-in. E-glass R65 SMC, 1-in. S-2 glass R65 SMC, and 2-in. S-2 glass R65 SMC. More explicitly, 1-in. E-glass R65 SMC indicates a material containing 65 weight percent 1-in.-long E-glass fibers that are randomly distributed, with the base material being or having been made from sheet molding compound (SMC).[2]

Mechanical Test Procedure and Results

All specimens of the same type were cut from the same panel by using a diamond impregnated sawblade and continuous coolant flow. Each was prismatic and 25.4 cm long, with a gage section 6.4 cm wide by 15.2 cm long, and was instrumented with a single 0-90 deg resistance foil strain gage pair. A wider gage section than is suggested by ASTM Test for Tensile Properties of Plastics (D 638) was used to avoid fibers in either case traversing the entire width. The specimens were tested in quasi-static tension at a constant rate of elongation. Clamping of the samples was accomplished by sandwiching the ends between pieces of 320 and 180 grit alumina oxide coated sand paper within hydraulic grips (Fig. 1).

Load and longitudinal and transverse strains were monitored continuously (Table 1). Details of particular tests and their significance are discussed in the sections dealing with the nondestructive evaluation. However, several distinct characteristics regarding the specimen failures are worth noting.

[2]Private communication with Dr. John F. Kay, Owens-Corning, Granville, Ohio 43023.

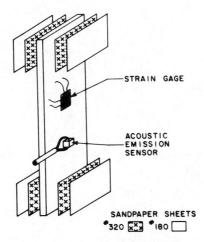

FIG. 1—*Specimen test configuration.*

TABLE 1—*Mechanical properties determined in uniaxial quasistatic tension.*

Material	Longitudinal Modulus		Poisson's Ratio	Ultimate Tensile Strength		Strain at Ultimate
	GPa	10^6 psi		MPa	10^3 psi	
S-2 Glass R65						
25.4 mm (1 in.)	18.6	2.7	0.31	264.8	38.4	0.018
50.8 mm (2 in.)	17.9	2.6	0.32	280.6	40.7	0.02
E-Glass R65						
25.4 mm (1 in.)	14.2	2.1	0.27	204.1	29.6	0.016
50.8 mm (2 in.)	17.2	2.5	0.32	244.8	35.5	0.017

Two types of final failure surfaces occurred and are shown in Fig. 2. One type of failure involved the severing of the specimen directly into two parts (Fig. 2*a*), with some axial splitting occurring along the midplane extending away from the plane of separation. The other type caused the specimens to end up in two pieces having a stepped appearance (Fig. 2*b*), with the axial splitting appearing to play a more significant part in the failure sequence in these cases. The failure stress that coincided with ultimate tensile strength appeared independent of the failure mode.

Examination of the NDE results provided considerable insight into these results.

NDE Procedure and Results

As has been indicated previously, consideration was limited to NDE techniques that would appear to have potential for application in industry.

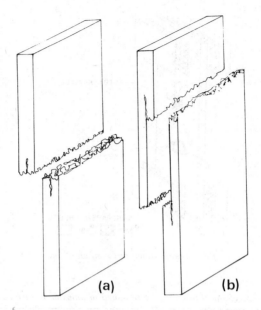

FIG. 2—*Typical failure patterns.* (a) *Transverse only.* (b) *Stepwise.*

The techniques considered included ultrasonic C-scanning, acoustic emission monitoring, X-ray radiography (without opaque penetrants, some of which are extremely toxic), and vibrothermography.

Ultrasonic C-Scanning

A conventional ultrasonic C-scan immersion system was used that consisted of a Sperry UM721 Reflectoscope and Automation US 450 Laboratory Scanner, along with a 10-MHz focused transducer. The specimens were interrogated through their thickness. In order to obtain an integrated estimate of any internal inhomogeneities, and because of limitations in spatial resolution caused by the relatively small thickness of the specimens, the scanning was performed as follows. Firstly, the specimens were positioned approximately 0.75 cm above a polished reflecting plate. The transducer position was then varied so that the ultrasonic pulse echo signal from the top surface of the specimen was maximized. The amplitude of the ultrasonic pulse echo signal, which resulted from the sound wave traveling through the specimen, striking the reflecting plate, and returning back through the specimen, was then monitored. By doing this, relative differences in the material's ability to transmit the sound waves could be plotted.

At this point it is necessary to consider in a somewhat general context the significance of this type of information. First, one is justified in suggesting that the smallest detectable difference in a particular material property, in this case the ability to transmit ultrasound, results from an imperfection in the material's condition in that region. Next, the question of degree of imperfection may be addressed. However, an operating definition could be motivated by a number of different philosophies, but would be only valid regarding the property being evaluated; that is, from the information obtained here, inference regarding other properties (strength, density, etc.) must be based on additional information. With this in mind, previous experience suggests that nearly complete, if not complete, attenuation of the ultrasound occurs if an interior surface exists. This condition was considered as a starting point; that is, determination of areas where little if any sound was transmitted. This criterion yielded C-scans indicating a number of areas of imperfection in the as-received materials (Fig. 3).

With the procedure described previously, the edges of the specimens are not properly scanned. The dark regions between the specimens are a result of this and are somewhat compounded by the proximity of the specimens during the scan. This is due to problems involving geometry, including the ultrasonic beam width, focal length, and specimen thickness, which give rise

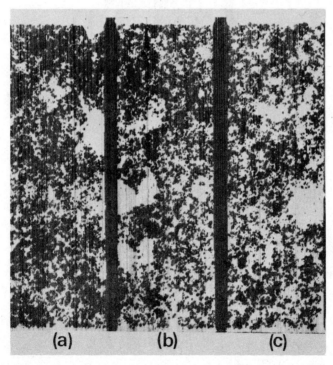

FIG. 3—*Typical ultrasonic C-scan taken before deformation, showing specimens cut from neighboring areas of the as-received panel. The arrow indicates the area shown in Fig. 4.*

to interference signals near the edge. It should be possible to minimize this problem with the appropriate equipment modifications. Nevertheless, from the C-scan it appeared that in some instances the region of imperfection extended to the edge of the specimen (Fig. 3a). Because of this, the edge was examined microscopically and a linear separation, located on the midplane of the specimen, was observed. Figure 4 is a photograph of a replica of the edge taken in the region indicated in Fig. 3a. It was not necessary to replicate the surface to photograph the discontinuity, but facilities in our laboratories for making hard copies of replicas made it expedient.

Prior to mechanical testing, every specimen was C-scanned in order to

FIG. 4—*Edge replica of planar imperfection for 3.5-mm-thick specimen.*

determine the presence and location of any imperfections of the type previously described. Now, since ultrasonic C-scanning by the procedure used in this study effectively monitors changes in ultrasonic attenuation, it would be reasonable to consider continuously monitoring changes in ultrasonic attenuation during the mechanical testing. This was not done, but it might have yielded continuous information regarding the development of damage in the area being monitored. However, ultrasonic C-scans in addition to the initial ones were made in several cases, two examples of which are shown in Figs. 5 and 6. Figure 5 shows the increase in size of two of the larger imperfections after the loading of the specimen was interrupted prior to failure, and Fig. 6 shows how the final failure occurs in the same location as the imperfection, indicated by the arrows. The location of the final failure in every specimen tested coincided with an area indicated by ultrasonic C-scanning to be an area of imperfection, when such areas were present.

EARLY UNLOAD   UNLOADED JUST
                               PRIOR TO FAILURE

FIG. 5—*Ultrasonic C-scans indicating growth of imperfections.*

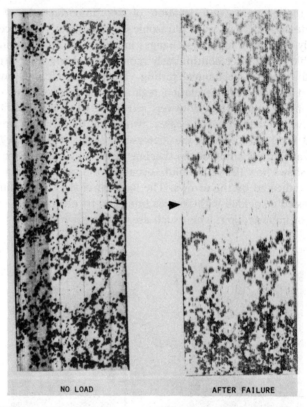

FIG. 6—*Ultrasonic C-scans showing correspondence of failure site (dark arrow) and imperfection location (light arrow).*

X-ray Radiography

A number of specimens were X-ray radiographed after the ultrasonic C-scanning and microscopic examination suggested imperfections were present. The system used was a Hewlett Packard 43805N X-ray System of the Faxitron series. Using Kodak Industrex M-5 film, an appropriate exposure time was found to be 4 min at 30 kV. Figure 7 is an example of a radiograph obtained in the manner prescribed without the use of X-ray opaque penetrants. Despite the fact that several large areas of imperfection were indicated by ultrasonic C-scanning and an area of planar separation was observed on one edge, no corresponding imperfection regarding X-ray transmission is visible. White areas of small spatial extent are evident and seem to suggest the use of this technique for examining the fiber distribution within the specimen. It should be noted that a planar defect oriented perpendicular to the X-ray beam would not be seen. Therefore, in examining objects

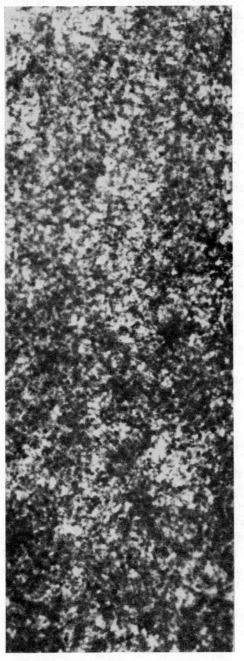

FIG. 7—*X-ray radiograph of specimen containing a large imperfection indicated by ultrasonic C-scan.*

containing defects that could be oriented in more than one direction, this technique might be useful. In addition, it should be noted that opaque penetrants would only be useful for parts with accessible edges that would be coincident with an imperfection.

Vibrothermography

Recent work by the Materials Response Group at the Virginia Polytechnic Institute and State University has shown that low-amplitude, steady-state mechanical vibrations of materials are capable of generating thermal patterns that reveal the nature of flaws in the material [1].[3] This technique is referred to as *vibrothermography* and involves the use of a mechanical shaker and a means of detecting the infrared radiation associated with the thermal changes. In this study a Wilcoxon Research F7 shaker and AGA Thermovision 680 were used. Each specimen was examined by clamping only one end to the vibration source and inertially loading it as a cantilever beam. Figure 8 is a schematic of the original 10-color picture obtained by photographing the real time image of the thermovision. It is possible to resolve differences in temperature of 0.1°C with the system. The white area indicated by the arrow is the hot region in the picture and corresponds to an area indicated by the ultrasonic C-scan. In many instances a correspondence between ultrasonic C-scanning and vibrothermography was obtained at particular frequencies of vibration. The significance of the thermal patterns developed at other frequencies is not clear.

Additional information that may be obtained by this technique is currently under further investigation in our laboratories. It should be pointed out that the technique is influenced by both the mechanical and thermal properties of the material.

Acoustic Emission Monitoring

After the initial series of nondestructive tests the specimens were mechanically deformed. During the deformation acoustic emission was monitored continuously. A Panametrics 100-kHz cross-coupled sensor was used along with a Panametric Preamplifier and a Tektronix 1A7A High Gain Differential Amplifier. Signals obtained were counted above a selectable threshold using a Hewlett-Packard 5326B Timer-Counter-DVM. The overall system gain was approximately 75 dB, with band pass filtering between 10 and 300 kHz. The typical counting threshold was 0.25 V.

Figures 9 to 12 show typical results obtained during the deformation of specimens for each material type tested. In each case the acoustic emission began to occur early in the deformation and progressively increased until

[3]The italic numbers in brackets refer to the list of references appended to this paper.

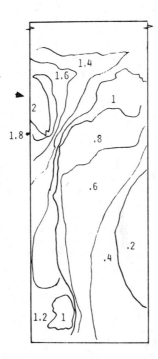

FIG. 8—*Schematic of a ten-color vibrothermographic pattern. Temperatures indicated are degrees Celsius above room temperature.*

failure. Tests were interrupted prior to failure in order to again take ultrasonic C-scans of the specimens. Such occasions were used in order to examine the Kaiser effect in these materials, which suggests that no acoustic emission occurs upon reloading of a specimen until after the previous maximum load is exceeded [2]. For the load values considered, the Kaiser effect was not observed (Fig. 13). During the initial loading and two subsequent reloadings the emission began shortly after the application of load, with little if any difference in the rate of emission observable. In addition, for this case failure occurred during the third loading at a value slightly less than the maximum value of load attained in the previous loading.

Such results suggest that acoustic emission activity is associated with the development of damage in these materials. Along with this it would seem that the presence of emission implies that the damage is in fact occurring and might continue to occur at a fixed load as long as the acoustic emissions persist. In the case shown, additional damage occurs at values below maximum load on reloading. McElroy [3] indicated that failure at a fixed load could occur after a time in fiberglass used in the construction of booms on aerial lift trucks. Such possibilities were not examined in this study.

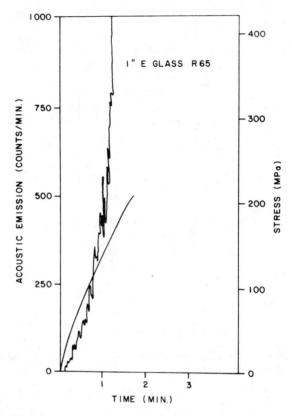

FIG. 9—*Typical acoustic emission activity for 25.4-mm (1-in.) E-glass R65.*

Discussion

Comparing the results of the NDE methods considered indicates the benefits that may be derived from considering more than one type of NDE data. For example, although in this study little if any information descriptive of the imperfections observed by ultrasonic C-scanning was obtained by the use of X-ray radiography, the latter technique allowed for the elimination of several possibilities which otherwise might have been considered. Further, the ultrasonic C-scanning appeared to provide a means of measuring the size of the imperfections, while the vibrothermography seemed to delineate some and not others. Next, in considering the effects of deformation on the observed imperfections by examining additional ultrasonic C-scans, it was possible to determine that the imperfections were in fact increasing in size. However, despite the failure patterns, which correlated extremely well with imperfections detected in the ultrasonic C-scans, mechanical performance

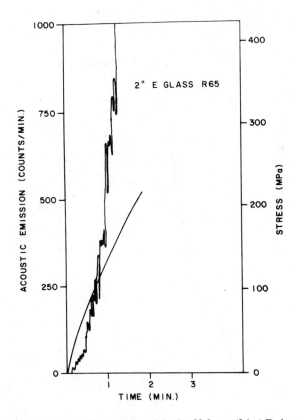

FIG.10—*Typical acoustic emission activity for 50.8-mm (2-in.) E-glass R65.*

was not seen to be affected by these imperfections. This is wholly consistent with the nature of the loading considered, since the imperfections were oriented parallel to the load direction. Being in this orientation and lying on the midplane of the specimen caused the imperfections to have little if any effect on the state of axial stress. The midplane splitting would be expected from the geometry of the imperfections, as a result of the Poisson effect acting on either side of the imperfection and the local stress concentration. Although the final failure stress did not appear to be influenced by these imperfections, if the state of stress were different—for example, as a result of flexure—these same imperfections should be very significant.

On the other hand, the lone passive technique—acoustic emission monitoring—suggests that damage begins to occur very early in the deformation of these materials. Damage progressed despite the specimen having been loaded previously.

Results of this preliminary study suggest a need to investigate more fully

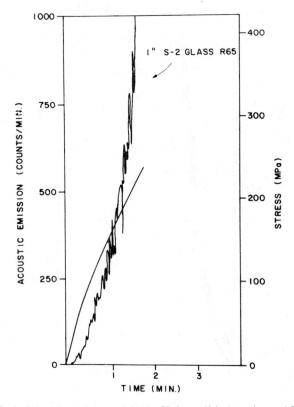

FIG.11—*Typical acoustic emission activity for 25.4-mm (1-in.) specimen of S-2 glass R65.*

the significance of the results obtained nondestructively in an effort to describe their relationship to the mechanical response of these materials. Due to the limited number of specimens tested and the singular nature of the mechanical test (that is, quasi-static tension), a complete description is impossible. Nevertheless, the information obtained seems consistent for the nondestructive techniques, the failure patterns, and mechanical results. These methods clearly provide valuable information not available through mechanical testing alone. In particular, the possible occurrence of damage at early stages of the deformation, indicated by the acoustic emission results, suggests that fatigue cycling at low loads may cause considerable damage; such a thought is not suggested by the mechanical test results.

Conclusions

1. Ultrasonic C-scanning detected planar imperfections.
2. Vibrothermographic imperfection indications tended to correlate well with ultrasonic C-scanning for large imperfections.

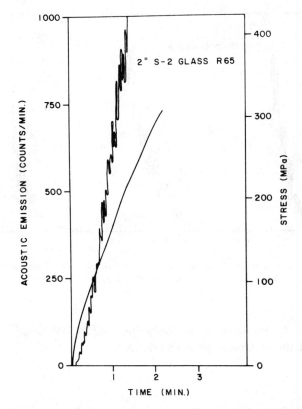

FIG.12—*Typical acoustic emission activity for 50.8-mm (2-in.) S-2 glass R65.*

3. X-ray radiography would appear to be useful for determining fiber distribution. This, however, was not substantiated in this study.

4. Acoustic emission activity began early in the deformation and increased toward failure.

5. The Kaiser effect of acoustic emission was not obeyed for the load values considered in this study.

6. The imperfections observed nondestructively appear to be responsible for the stairstep failure patterns observed.

7. The imperfections observed would be expected to behave as defects in situations involving shear stress on the midplane of the part. The behavior of these imperfections in such situations could more properly be examined by means of a three-point bend test.

Acknowledgments

This work was supported in part by the Lawrence Livermore Laboratories; Dr. S. V. Kulkarni was the technical monitor. The helpful suggestions of my

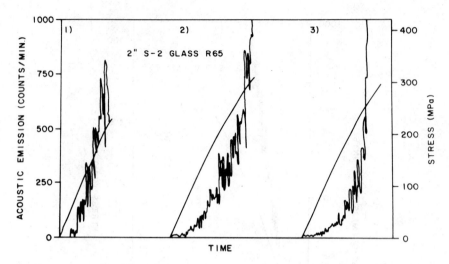

FIG.13—*Acoustic emission activity in 50.8-mm (2-in.) S-2 glass R65 during three consecutive loading and unloading cycles, demonstrating the absence of the Kaiser effect.*

colleagues E. G. Henneke, K. L. Reifsnider, and W. W. Stinchcomb of the Materials Response Group are acknowledged.

References

[*1*] Henneke, II, E. G., Reifsnider K. L., and Stinchcomb, W. W., *Journal of Metals*, Vol. 31, No. 9, Sept. 1979, pp. 11–15.

[*2*] Kaiser, J., *Arkiv für das Eisenbuttenwesen*, Vol. 24, 1953, p. 43.

[*3*] McElroy, J. W., "On-Board Acoustic Emission Monitoring of Fiberglass Boom Aerial Lift Trucks," SAE Technical Paper Series 800070, Society of Automotive Engineers, Warrendale, Pa., 1980.

*D. E. Walrath,[1] D. F. Adams,[1] D. A. Riegner,[2]
and B. A. Sanders[2]*

Mechanical Behavior of Three Sheet Molding Compounds

REFERENCE: Walrath, D. E., Adams, D. F., Riegner, D. A., and Sanders, B. A., **"Mechanical Behavior of Three Sheet Molding Compounds,"** *Short Fiber Reinforced Composite Materials, ASTM STP 772,* B. A. Sanders, Ed., American Society for Testing and Materials, 1982, pp. 113-132.

ABSTRACT: A test program was conducted to measure various mechanical properties of three sheet molding compounds, SMC-R25, SMC-R30, and SMC-R65. Static tension tests were conducted in ambient and elevated temperature environments on both dry and moisture preconditioned specimens. Compression, shear, flexure, and impact tests were also conducted; these were performed at ambient temperatures on dry specimens.

KEY WORDS: composite materials, sheet molding compounds, mechanical properties, elevated temperature, fatigue, impact

One approach toward building more fuel-efficient automobiles is to reduce overall vehicle weight. This can be accomplished by use of light-weight E-glass reinforced polyester sheet molding compounds (SMC). In order to make optimum use of SMC materials, the mechanical behavior of these materials under various loading and environmental conditions must be measured. During the course of the work discussed in this paper, three sheet molding compounds, designated SMC-R25, SMC-R30, and SMC-R65, were subjected to a variety of mechanical tests, some of which were carried out in elevated temperature, humid environments.

Static tension tests were performed in ambient and elevated temperature test environments on both dry and wet preconditioned test coupons. Wet preconditioning was carried out at 65°C, 98 percent relative humidity. Tension specimens were instrumented with a strain gage extensometer; therefore, both strength and elastic modulus measurements were recorded.

[1]Department of Mechanical Engineering, University of Wyoming, Laramie, Wyo. 82071.
[2]Manufacturing Development, General Motors Technical Center, Warren, Mich. 48081.

Similar behavior was observed for all three materials, with strength and modulus decreasing as a function of test temperature. Moisture absorption also caused losses in strength and elastic modulus, but further reductions due to a combination of test temperature and moisture absorption were not observed.

Compression tests were conducted at ambient and elevated test temperatures, but only for dry test specimens. Due to limited space within the test fixture, no modulus measurements were made. Compressive strength as a function of temperature was observed to be quite similar to the tension test results.

Shear strengths of the three materials were measured using a short beam shear test configuration. These tests were performed at room tempreature on dry test specimens. Bending as well as shear failures were observed during this testing, indicating that the short beam test method is not an ideal shear strength test method for this class of materials. Values obtained for the shear strengths do represent a lower bound on that strength, however.

Both three-point and four-point flexure tests were conducted for dry specimens at ambient temperatures. Strength and elastic modulus results of these tests exhibited close similarity between the two flexure test methods used, but were greater than strength and elastic modulus values obtained with the tension test method. This is not uncommon behavior for composites, however, pointing out the complexity of the stress state within a flexure sample during testing.

Both tension and Charpy impact tests were conducted at ambient temperature on dry test specimens. Tensile impact strengths were found to be comparable to static test values. Energy absorption during impact was also calculated. Various Charpy tests were conducted on specimens oriented in different directions. Differences in strength and energy absorption were observed depending on the orientation of the specimens.

Materials Description

Three E-glass fiber-reinforced polyester sheet molding compounds were tested. These materials, designated SMC-R25, SMC-R30, and SMC-R65, contain 25, 30, and 65 percent by weight of 25.4-mm (1-in.)-long, chopped E-glass fiber, respectively. Specific material formulations are shown in Table 1 [1].[3] Each of the SMC materials was nominally 2.5 mm (0.1 in.) thick. In order to eliminate anisotropic behavior as a test variable, most testing was performed in one primary orientation for each material. A limited number of tests were also conducted on specimens in the perpendicular or transverse direction, however, to detect any anisotropic behavior. Arbitrary primary directions were chosen for the SMC-R30 and SMC-R65 materials, as they ex-

---

[3]The italic numbers in brackets refer to the list of references appended to this paper.

TABLE 1—*SMC material formulations.*

Material Designation	Constituent	Function	Weight Percent
SMC-R25	E-glass (OCF 951 AB)[a]	fiber	25.0
	polyester (OCF E-920-1)	resin	29.4
	calcium carbonate	filler	41.8
	zinc stearate	internal release	1.1
	tertiary butyl perbenzoate	catalyst	0.3
	magnesium hydroxide	thickener	1.5
	mapico black	pigment	0.8
SMC-R30	E-glass (OCF-495)	fiber	29.0
	polyester	resin	19.9
	calcium carbonate balance	filler	41.0
	balance	thickener etc.	11.1
SMC-R65	E-glass (PPG 518)[b]	fiber	65.0
	polyester (PPG 50271)	rigid resin	16.0
	polyester (PPG 50161)	flexible resin	16.0
	balance	thickener etc.	3.0

[a]OCF = Owens-Corning Fiberglas.
[b]PPG = PPG Industries.

hibited no visual indication of preferred fiber orientation. However, the SMC-R25 did appear to exhibit a preferred fiber orientation; this direction was chosen as the primary direction.

Static Tension Tests

A total of 180 tension tests were performed, 60 on each material, at four temperatures and two preconditioning environments, for two different specimen orientations. Tension tests were performed at room temperature, 60°, 90°, and 150°C. Moisture preconditioned (wet) test specimens were exposed to a 65°C, 98 percent relative humidity environment for approximately eight weeks. Elevated temperature testing was performed in a simple clamshell heating unit. Rapid surface dryout of tensile specimens during elevated temperature testing can cause large moisture gradients, and therefore large stress gradients, within a test specimen. A number of tension tests on moisture preconditioned specimens were therefore also conducted at 150°C under pressure. In order to limit dryout effects, a saturated steam environment was maintained around these test specimens, requiring a pressurized test chamber. Pressure chamber specimens were preconditioned for 12 weeks instead of 8 weeks, so open air 150°C tests were repeated for other specimens also preconditioned for 12 weeks. A limited number of room-temperature tension tests were performed on wet and dry specimens cut from the transverse orientation. This was done to detect any anisotropy of the three materials. All tension specimens, except the pressure chamber specimens, were instrumented with a strain gage extensometer to measure elastic modulus. Testing was conducted according to ASTM Test for Tensile Properties of Plastics (D 638), using a crosshead speed of 2 mm/min.

Averaged tensile strength and modulus results are presented in Figs. 1 to 6. As can be seen in Figs. 1 and 2, SMC-R25 appears to be quite anisotropic, showing a 36-MPa tensile strength at room temperature in the transverse (90 deg) orientation and an 117-MPa tensile strength in the primary orientation. This anisotropy was expected as a definite fiber orientation was visually evident in the SMC-R25 panels, this direction being chosen as the primary specimen orientation. The strengths and moduli decreased with both test

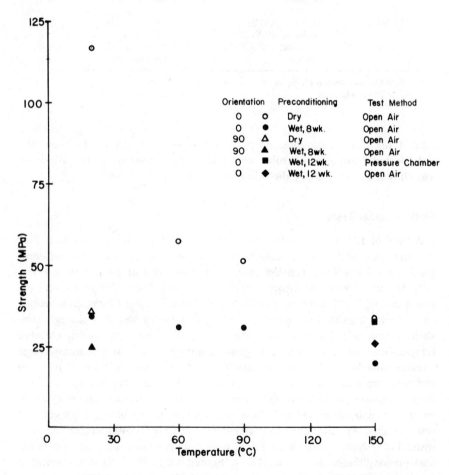

FIG. 1—*Average tensile strength versus temperature for SMC-R25.*

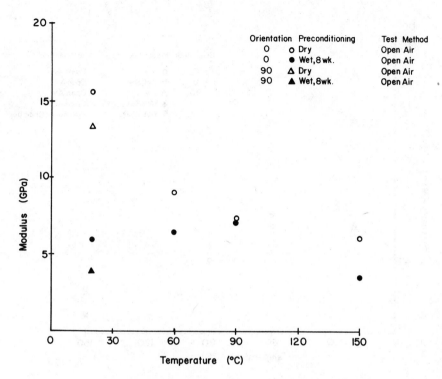

FIG. 2—*Average tensile modulus versus temperature for SMC-R25.*

temperature and moisture absorption (Figs. 1 and 2). The open circles in Fig. 1 indicate the average tensile strength of SMC-R25 as a function of test temperature. The tensile strength of the dry specimens dropped from 117 MPa at room temperature to 33 MPa at a test temperature of 150°C. Temperature had little effect on the strength of the moisture preconditioned specimens, however, as indicated by the solid circles in Fig. 1. The average strengths of wet tension specimens ranged from 34 MPa at room temperature to 19 MPa at 150°C. It would appear that absorbed moisture has a much greater effect on the tensile strength of SMC-R25 than does temperature; the combined effect of temperature and absorbed moisture is not much greater than that of absorbed moisture alone. The effects of temperature and moisture on the elastic modulus, as shown in Fig. 2, follow the same trends as for the tensile strengths shown in Fig. 1.

Five tension tests were conducted on 12-week preconditioned specimens using a pressure vessel environment at 150°C. This was done to eliminate any effect of surface moisture desorption during these relatively high temperature tests. A solid square symbol is used in Fig. 1 to indicate the average 32-MPa tensile strength obtained. An extensometer was not used during these tests;

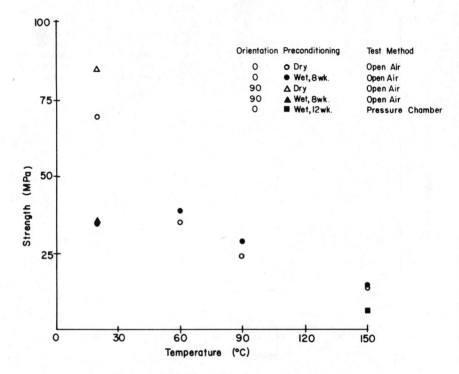

FIG. 3—*Average tensile strength versus temperature for SMC-R30.*

therefore, no corresponding modulus value is shown in Fig. 2. A series of five tension tests were also conducted in the clam shell oven at 150°C on the 12-week preconditioned specimens, for direct comparison with the pressure chamber test results. This average strength of 26 MPa is shown as a solid diamond symbol in Fig. 1. As can be seen, the strength obtained in the pressure chamber tests is greater than those values obtained in open air tests, which indicates that surface dryout may be significant.

Thus, from Figs. 1 and 2, it can be seen that the strength and modulus of the primary orientation specimens dropped rapidly as the test temperature increased. Moisture absorption caused a reduction in strength and modulus which was fairly constant independent of the test temperature. Moisture absorption also caused a reduction in strength and modulus of the transverse specimens, as indicated by the open and solid triangles.

Average tensile strengths and elastic moduli for the SMC-R30 material are plotted in Figs. 3 and 4, respectively. The SMC-R30 was only slightly anisotropic, with the stronger direction being that which was defined as the transverse (90 deg) orientation. Similar to the SMC-R25 tensile results, strength and modulus dropped rapidly with increasing test temperature, and

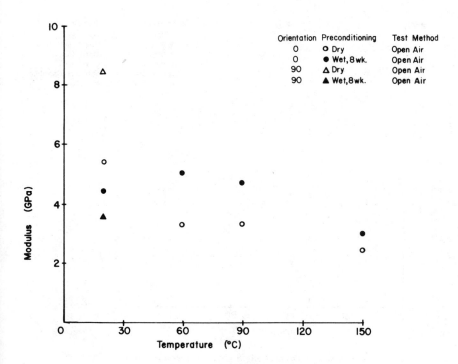

FIG. 4—*Average tensile modulus versus temperature for SMC-R30.*

with moisture absorption. The wet tension specimens tested at elevated temperatures had higher strengths and higher elastic moduli than wet specimens tested at room temperature. However, this difference is within the data scatter. The average tensile strength of dry SMC-R30 specimens ranged from 69 MPa at room temperature to 14 MPa at 150°C, while the corresponding average elastic moduli were 4.8 GPa and 2.5 GPa. The wet specimens exhibited strengths of 36 and 15 MPa, and elastic moduli of 4.4 and 3.0 GPa, at room temperature and 150°C, respectively. The pressure chamber strength results were slightly less than values obtained from elevated temperature test results; however, the preconditioning times were different.

Test results for the SMC-R65 are plotted in Figs. 5 and 6. Again, the trend of decreasing strength and modulus with increasing temperature is seen. Moisture absorption caused a degradation of strength which became increasingly more severe with increasing test temperature. The elastic moduli of the wet specimens remained fairly constant with temperature, except for the specimens tested at 60°C. Although this point seems somewhat high relative to the trend, the scatter in the results of these five tests was quite low, suggesting reliability of the data.

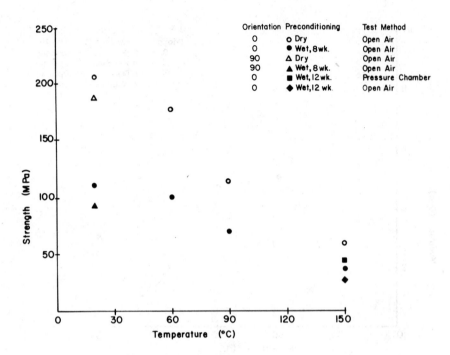

FIG. 5—*Average tensile strength versus temperature for SMC-R65.*

To summarize the tension test results for the SMC-R25, SMC-R30, and SMC-R65 materials as presented in Figs. 1 to 6, tensile strength decreased with both temperature and moisture, but the combined effects of elevated temperature and moisture were not significant. Although some anomalies are present, elastic modulus data appeared to follow the same trends as the strength values. Pressure chamber tests exhibited some strength differences when compared with open air tests, which indicates that surface dryout of moisture-preconditioned specimens may be significant during elevated temperature testing.

Diffusion Behavior

A limited number of specimens were monitored for weight gain during the moisture preconditioning of the hydrothermal tension specimens. Initially, all three materials exhibited Fickian diffusion behavior, and diffusion coefficients were calculated using the method described in Ref 2. These values ranged from 5.4×10^{-12} m²/s for SMC-R25 to 2.1×10^{-12} m²/s for SMC-R65. Too few tests were conducted to attach absolute quantitative significance to these numbers. However, these values indicate diffusion coefficients

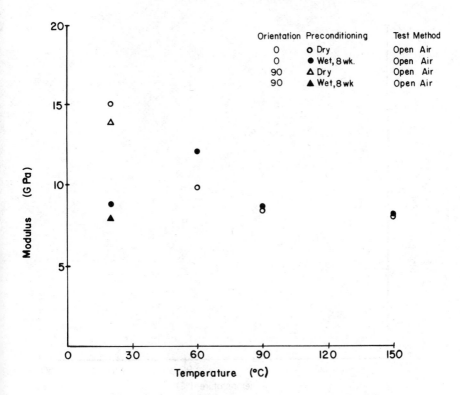

FIG. 6—*Average tensile modulus versus temperature for SMC-R65.*

that are 30 to 40 times larger than comparable values obtained for epoxy matrix composites. Also, during the preconditioning the SMC-R25 and SMC-R65 specimens began to blister and lose weight, which suggests that some chemical change was taking place.

Static Compression Tests

A total of 60 compression tests were conducted using the fixture and methods described by ASTM Test for Compressive Properties of Unidirectional or Crossply Fiber-Resin Composites (D 3410). Five specimens from each material were tested at room temperature, 60, 90, and 150°C. These specimens were straight-sided with glass fabric/epoxy tabs bonded to each end. Due to space constraints, an extensometer could not be used in these tests. All compression tests were conducted on dry, primary orientation specimens. Average compressive strength values for the three materials are plotted in Figs. 7 to 9.

Examining the compressive results for SMC-R25 shown in Fig. 7, we see

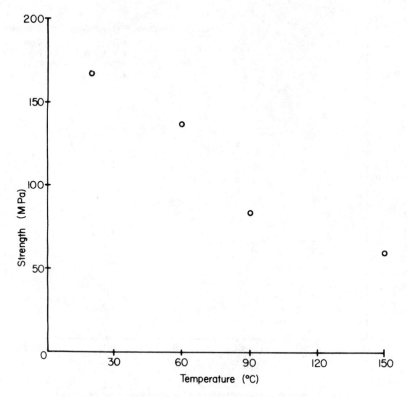

FIG. 7—*Average compressive strength versus temperature for SMC-R25.*

that compressive strength decreased with increasing test temperature, as might be expected. Average strength values for SMC-R25 ranged from 168 MPa at room temperature to 61 MPa at 150°C. Comparing Fig. 7 with Fig. 1, we note that the tensile and compressive strengths decreased in much the same manner, although the compressive strengths were higher in magnitude.

The same trend in compressive strength data existed for the SMC-R30 material (Fig. 8); again, strength dropped rapidly with increasing test temperature. Comparing Fig. 8 with Fig. 3, we see that the tensile and compressive strengths as a function of temperature were very similar, even in magnitude.

Compressive strength results for the SMC-R65 are plotted in Fig. 9. The same decreasing strength versus temperature trend observed for the two other materials was again indicated, with strengths ranging from 241 MPa at room temperature to 31 MPa at 150°C. The SMC-R65 compressive strength curve does not appear to be of the same shape as the tensile strength versus

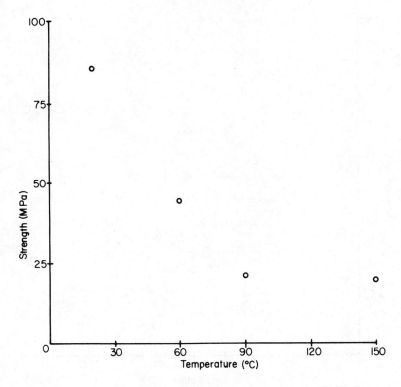

FIG. 8—*Average compressive strength versus temperature for SMC-R30.*

temperature curve shown in Fig. 5; compressive strength seems to drop much more rapidly with increasing temperature than does tensile strength.

Shear Tests

Short beam shear tests were performed in accordance with ASTM Test for Apparent Interlaminar Shear Strength of Parallel Fiber Composites by Short Beam Method (D 2344) at room temperature on dry, primary orientation specimens, using a crosshead rate of 0.2 mm/min. Six tests were conducted on each of the three materials. Shear modulus cannot be determined from this test. Short beam shear strengths were calculated for each of the materials; average values are listed in Table 2. There is some doubt, however, that the specimens actually failed in shear. Close examination of the specimens, especially those of SMC-R25 and SMC-R30, revealed cracks across the bottom surface, indicating tensile failure due to bending rather than by shear. These cracks were also present in some of the failed SMC-R65 shear specimens, indicating they may have failed in bending also, while others

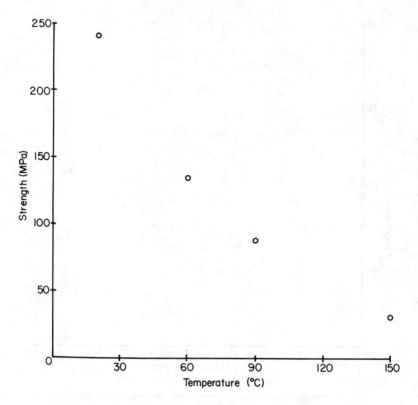

FIG. 9—*Average compressive strength versus temperature for SMC-R65.*

clearly failed in shear. The short beam shear test does not appear to give reliable shear strength results for these types of composite materials; however, the values presented in Table 2 do represent a lower bound on the shear strength. A newer test, the Iosipescu shear test or the Asymmetric Four-Point Bending (AFPB) shear test is currently being developed at the University of Wyoming for characterization of the shear properties of SMC materials [3,4].

Static Flexure Tests

Both three-point and four-point static flexure tests were conducted at room temperature on dry, primary orientation test specimens. Five four-point and six three-point tests were run for each material, a total of 33 separate tests. These experiments were conducted in accordance with ASTM Tests for Flexural Properties of Plastics and Electrical Insulating Materials (D 790), using

TABLE 2—*Average short beam shear results.*[a]

Material	Strength	
	MPa	(ksi)
SMC-R25	30	(4.3)
SMC-R30	32	(4.6)
SMC-R65	45	(6.5)

[a]Tests were conducted on dry specimens at room temperature.

a crosshead speed of 1 mm/min for three-point flexure and 5 mm/min for four-point flexure.

Results of the flexure tests are presented in Table 3. Strengths and elastic moduli did not differ significantly between the two different types of flexure loading, except for the SMC-R65 material. Strengths obtained by flexural loading did tend to differ from strengths obtained from pure tensile loading or pure compressive loading.

Impact Tests

Two types of instrumented impact tests were conducted on the three materials: conventional Charpy impact and uniaxial tensile impact. A total of 58 specimens were tested in tensile impact on an instrumented, pendulum-type (Satec) impact testing machine, using a special gripping fixture developed at the University of Wyoming [5]. All tests were conducted at room temperature, using both primary and transverse orientation specimens. Two different impact velocities, 3.35 and 5.18 m/s (11 and 17 ft/s) were used for

TABLE 3—*Average static flexure test results.*[a]

Material	Loading	Strength		Modulus	
		MPa	(ksi)	GPa	(Msi)[b]
SMC-R25	3-point	220	(31.9)	4.8	(0.69)
	4-point	186	(26.9)	5.9	(0.86)
SMC-R30	3-point	131	(18.9)	3.5	(0.51)
	4-point	134	(19.4)	4.3	(0.63)
SMC-R65	3-point	403	(58.4)	5.7	(0.82)
	4-point	298	(43.2)	6.9	(1.00)

[a]Tests were conducted on dry specimens at room temperature.
[b]Msi $= 10^6$ psi.

these tests. Tensile impact specimens were straight-sided, 9.5 mm wide and 127 mm long (0.375 in. by 5.0 in.), with 25-mm (1-in.) glass fabric/epoxy tabs bonded to each end. The tensile grip fixture was instrumented with strain gages to provide real time load data, recorded and processed by a Hewlett-Packard minicomputer (HP 2100) data acquisition system.

Approximately 25 Charpy impact tests were performed on each of the three SMC materials. The standard Charpy impact specimen of ASTM Methods for Notched Bar Impact Testing of Metallic Materials (E 23) is 50 mm long and 10 mm square in cross section (1.9 by 0.39 by 0.39 in.), with a 45-deg notch, 2 mm deep, located on the tension side at the center of the span. In order to conform as closely as possible to the ASTM standard, the SMC Charpy specimens were fabricated by first laminating three thicknesses of each panel together with an epoxy adhesive, in order to obtain a specimen having a thickness as close as possible to the desired 10 mm. Specimens were then cut to the above-stated dimensions, but were not notched. Previous studies [6] have indicated that composite materials are notch-insensitive. The majority of tests were conducted with the impact direction perpendicular to the plane of lamination. However, a limited number of tests were performed with the direction of impact parallel to the lamination plane. Tests were performed with the primary material direction both parallel and perpendicular to the long axis of the test specimen. These tests were also conducted at room temperature on dry specimens, using impact velocities of 3.35 and 5.18 m/s (11 and 17 ft/s). The Charpy loading tup was instrumented with strain gages in order to allow measurement of load during impact, this information being recorded by the same minicomputer data acquisition system used for tensile impact tests.

Average properties for the three SMC materials tested in instrumented tensile impact are listed in Table 4. Impact energy values were calculated by numerically integrating the area under the impact force versus time curve, and accounting for the impact velocity. Normalized peak force and the corresponding normalized peak energy absorption were calculated by dividing peak values by the cross-sectional area of each specimen. Therefore normalized peak force values are merely the peak stress, or impact strength, of the material.

A representative tensile impact data plot is shown in Fig. 10. As can be seen, the force increased very rapidly to a peak value, at which time failure occurred. Only a short time was required to complete the failure process, less than one millisecond. Figure 10 shows data for an SMC-R65 specimen tested at a 5.18 m/s (17 ft/s) impact velocity. Curve shapes for the other two materials were very similar.

To summarize the data presented in Table 4, the impact strengths of the three SMC materials closely matched the static strengths indicated in Figs. 1, 3, and 5. Impact strength variations between primary and secondary orientation specimens existed in some instances, as they did in the static data. The

TABLE 4—Tensile impact results.[a]

Material	Test Direction, deg	Impact Velocity, m/s	Peak Force, kN	Energy at Peak Force, J	Total Energy, J	Normalized Peak Force, MPa	Normalized Energy at Peak Force, kJ/m^2	Normalized Total Energy, kJ/m^2
SMC-R25	0	3.35	4.1	5.2	6.9	113	146	192
	0	5.18	3.3	2.4	3.2	99	68	90
	90	5.18	1.8	0.9	1.5	49	39	25
SMC-R30	0	3.35	2.1	1.6	2.2	65	49	69
	0	5.18	2.1	1.5	1.9	65	46	58
	90	5.18	2.7	1.9	2.4	81	59	73
SMC-R65	0	3.35	7.4	13.9	15.6	224	317	354
	0	5.18	5.5	8.6	9.9	181	265	308
	90	5.18	9.1	13.6	18.6	202	303	414

[a]Tests were conducted on dry specimens at room temperature.

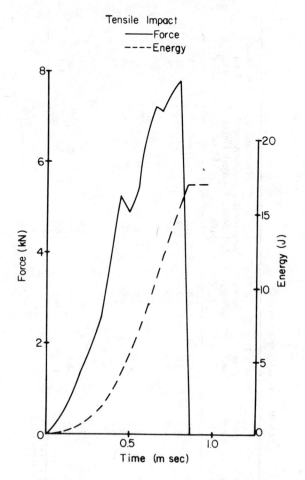

FIG. 10—*Sample tensile impact plot for SMC-R65.*

testing rate appears to have had little effect on the strength values. Energy values followed the same trends as the impact strengths. It should be noted that approximately 25 percent of the total energy absorbed by the specimen occurred after peak force had occurred.

Averaged results from the Charpy impact tests are shown in Table 5. Normalizing was accomplished by dividing by the cross-sectional area of each specimen. Although this normalization is not very meaningful from a mechanics or stress analysis viewpoint, in that it does not result in a valid stress indication, it is common practice and does serve to account for dimensional variations from specimen to specimen. Test direction refers to the direction parallel to the long axis of the test specimen. Impact direction is

TABLE 5—*Charpy impact results.*[a]

Material	Test Direction	Impact Direction	Impact Velocity, m/s	Energy at Peak Force, kN	Peak Force, J	Total Energy, J	Normalized Peak Force, MPa	Normalized Energy at Peak Force, kJ/m²	Normalized Total Energy, kJ/m²
SMC-R25	x	z	3.35	3.9	1.6	9.1	38	15	89
	x	z	5.18	3.8	1.4	6.4	36	13	63
	x	y	5.18	4.1	1.9	5.9	41	20	60
	y	z	5.18	2.8	0.9	3.7	27	9	35
SMC-R30	x	z	3.35	3.0	1.9	10.3	32	21	111
	x	z	5.18	3.9	1.8	5.5	40	19	57
	x	y	5.18	4.1	2.2	4.6	42	23	48
	y	z	5.18	2.7	1.1	4.4	28	12	49
SMC-R65	x	z	3.35	5.4	3.5	23.0	55	36	236
	x	z	5.18	4.2	2.7	13.0	42	27	131
	x	y	5.18	8.0	4.2	10.4	86	45	112
	y	z	5.18	5.5	4.0	20.4	57	42	209

[a]Tests were conducted on dry specimens at room temperature.

parallel to the impact striker (tup) velocity (that is, perpendicular to the test direction). Coordinates x, y, and z were chosen to represent the specimens as follows:

x = primary fiber direction (0 deg)
y = secondary (perpendicular) direction (90 deg)
z = perpendicular to the plane of the plate

For example, the first entry in Table 5 is for the SMC-R25 specimens which were tested with the primary orientation (x) parallel to the long axis of the specimen. The direction of impact was perpendicular to the lamination plane, (that is, parallel to the z-axis of the specimen).

A representative Charpy impact data plot is shown in Fig. 11 for an SMC-R65 specimen tested with an impact velocity of 5.18 m/s. Note the relatively long time period over which failure occurred. Several failure mechanisms occur simultaneously, including delamination, shear, and tensile failures, making Charpy impact tests of composites extremely difficult to analyze compared with the relatively simple tensile impact tests.

As can be seen in Table 5, variations in specimen orientation produced variation in strength and absorbed energy. As was mentioned earlier, the listed strengths are not very meaningful from a mechanics viewpoint and therefore cannot be directly compared to tension or compression results. Charpy strength results are useful for comparing the three materials, however. Note that the energy at peak force is far less than the total energy, indicating most of the energy was absorbed after the peak. This is different from the pure tensile impact results, where much of the energy was absorbed before the peak force.

Conclusions

Comparing the three materials tested in this program, a number of similarities and differences become obvious. The 25 weight percent fiber SMC-R25 was highly anisotropic, as evidenced by a visually detectable preferred fiber orientation as well as considerable difference between primary and transverse properties. Little anisotropy was detected for the slightly greater (30 percent) fiber content SMC-R30 or for SMC-R65. All three materials showed degradation in tensile and compressive material properties at elevated temperatures, and tensile property degradation due to moisture absorption. A combination of absorbed moisture and elevated temperature did not seem to cause severe additional degradation, however.

All three materials were shown to readily absorb moisture, causing the previously mentioned property changes. Indications of chemical change—that is, weight loss and surface blistering during the 65°C, 98 percent relative humidity preconditioning—were noted and should be further studied to determine the significance in terms of changes in mechanical properties.

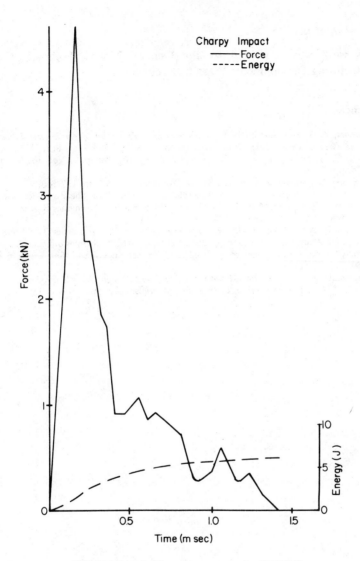

FIG. 11—*Sample Charpy impact plot for SMC-R65.*

Tensile impact tests results indicated few differences from static tensile data. Loading rate did not appear to have a significant effect. Approximately 25 percent of the total impact energy was shown to be absorbed after peak load, a fact to be considered in designing energy-absorbing structures. Charpy impact tests were useful for comparing the different materials, but the stress state within these specimens is far too complex to permit the generation of useful quantitative data.

Acknowledgments

This work was sponsored by the General Motors Technical Center, Manufacturing Development, Warren, Michigan.

References

[1] Sanders, B. A. and Heimbuch, R. A., *Engineering Properties of Automotive Fiber Reinforced Plastics,* Report MD-77-020, General Motors Corp., Warren, Mich., 1977.
[2] Shen, C. H. and Springer, G. S., *Journal of Composite Materials,* Vol. 10, No. 1, Jan. 1976, pp. 2–20.
[3] Iosipescu, N., *Journal of Materials,* Vol. 2, No. 3, Sept. 1967, pp. 537–566.
[4] Slepetz, J. M., Zagaeski, T. F., and Novello, R. F., "In-Plane Shear Test for Composite Materials," Report AMMRC TR 78-30, Army Materials and Mechanics Research Center, Watertown, Mass., July 1978.
[5] Benson, R. A. and Adams, D. F., "An Instrumented Tensile Impact Test Method for Composite Materials," *Proceedings,* 25th National SAMPE Symposium and Exhibition, Society for the Advancement of Materials and Process Engineering, San Diego, 6–8 May 1980.
[6] Adams, D. F. in *Composite Materials: Testing and Design (Fourth Conference), ASTM STP 617,* American Society for Testing and Materials, 1977, pp. 409–426.

R. F. Gibson,[1] Anna Yau,[1] and D. A. Riegner[2]

Vibration Characteristics of Automotive Composite Materials

REFERENCE: Gibson, R. F., Yau, Anna, and Riegner, D. A., "**Vibration Characteristics of Automotive Composite Materials,**" *Short Fiber Reinforced Composite Materials, ASTM STP 772*, B. A. Sanders, Ed., American Society for Testing and Materials, 1982, pp. 133-150.

ABSTRACT: This paper describes measurements of dynamic stiffness and internal damping (that is, the complex moduli) of automotive fiber-reinforced plastics over the frequency range 10 to 1000 Hz at maximum strain amplitudes up to 0.0017. Three types of chopped E-glass fiber-reinforced polyester composites and two types of hybrid chopped/continuous fiber-reinforced polyester composites are tested, along with neat resin samples. Complex moduli are measured by using a forced flexural vibration technique. Materials having the greatest stiffness are generally found to have the lowest damping, and vice versa. The complex moduli of all materials are found to be essentially independent of frequency and amplitude within the ranges investigated. Damping in the composites is at least one order of magnitude greater than the damping in an aluminum calibration specimen.

KEY WORDS: composite materials, damping, dynamic stiffness, vibration, reinforced plastics

Until recently, composite materials research has been centered on continuous fiber-reinforced plastics for aerospace applications. However, current requirements in the automobile industry have resulted in the development of relatively inexpensive chopped fiber and hybrid chopped/continuous fiber composites. The static mechanical properties of these new materials have been characterized [1],[3] but up to now little was known of their dynamic properties. This paper summarizes the results of the first phase of an ex-

[1]Associate Professor, Engineering Science and Mechanical Engineering Depts., and Research Assistant, Mechanical Engineering Dept., respectively, University of Idaho, Moscow, Idaho 83843.

[2]Project Engineer, Plastics Materials Characterization, General Motors Manufacturing Development, GM Technical Center, Warren, Mich. 48090.

[3]The italic numbers in brackets refer to the list of references appended to this paper.

perimental program for the generation of such data. The purpose of this first phase was to generate baseline data for small amplitude vibration at various frequencies under room temperature and humidity conditions. The second phase, which involved the investigation of the effects of various environmental conditions on the dynamic properties, is discussed in a separate paper [2].

Two categories of E-glass/polyester composites were tested: (1) chopped random fiber reinforced PPG SMC-R25, PPG SMC-R65, and OCF SMC-R25, and (2) hybrid chopped/continuous fiber reinforced PPG XMC-3 and OCF C20/R30.[4] General descriptions of these materials are given in Table 1. In addition, neat resin for the PPG SMC-R25 and PPG SMC-R65 was also tested.

In this paper, dynamic stiffness and internal damping are conveniently expressed in terms of the complex modulus $E*$:

$$E* = E' + iE'' = E'(1 + i\eta) \tag{1}$$

where

$E' =$ storage modulus,
$E'' =$ loss modulus,
$\eta =$ loss factor $= E''/E'$, and
$i = \sqrt{-1}$.

The complex modulus generally depends on frequency and temperature, but only frequency dependence is studied here; temperature dependence is studied in the second phase of the program [2]. Complex modulus notation makes it possible to convert an elastic analysis to an anelastic one simply by converting to complex variables. It should be mentioned, however, that the complex modulus notation only has meaning for sinusoidal vibration of

TABLE 1—Description of materials tested.

| | Weight Percentages of Constituents | | |
Material	Chopped E-Glass Fibers	Continuous E-Glass Fibers	Polyester Resin, Fillers, etc.
PPG SMC-R25[a]	25	0	75
PPG SMC-R65[a]	65	0	35
PPG XMC-3[a]	25	50 (± 7.5 deg, x-pattern)	25
OCF SMC-R25[b]	25	0	75
OCF C20/R30[b]	30	20 (aligned)	50

[a]Manufactured by PPG Industries, Fiber Glass Division, Pittsburgh, Pa. 15222.
[b]Manufactured by Owens-Corning Fiberglas Corporation, Toledo, Ohio 43659.

[4]The prefix PPG denotes materials manufactured by PPG Industries of Pittsburgh, Pa.; OCF denotes materials manufactured by Owens-Corning Fiberglas Corporation, Toledo, Ohio.

linear anelastic materials [3]. The linearity restriction means that stiffness and damping must be independent of amplitude, and this was found to be true for all the materials tested.

A resonant dwell technique, based on forced flexural vibration of cantilever beam specimens, was employed for complex modulus measurements [4]. This method was selected because it offers precise control over both the frequency and amplitude of vibration. The flexural mode was used because it is the most common mode of structural vibration. The complex modulus of a material specimen was found from measurements of resonant frequency, input acceleration, bending strain, specimen dimensions, and an assumed mode shape. Strain and acceleration signals were used to form a Lissajous pattern on an oscilloscope screen. Frequency effects were determined by testing specimens of various lengths up through the fifth mode of flexural vibration.

Static Flexure Tests

Dynamic stiffness (storage modulus) data, when extrapolated to very low frequency, should agree with static stiffness data. Thus static stiffness can provide a check on the storage modulus. Static stiffness was determined by using a four-point flexure test. Specimen dimensions of 127 by 25 by 3 mm (5 by 1 by ⅛ in.) and the support span of 102 mm (4 in.) were selected in accordance with ASTM Tests for Flexural Properties of Plastics and Electrical Insulating Materials (D790). A different loading span of 50.8 mm (2 in.) was used instead of the recommended loading span of 33.8 mm (1.33 in.) in order to provide sufficient clearance for a strain gage to be mounted at midspan. The static flexural modulus was found from the equation

$$E = \frac{3(L - a)}{2bt^2} \left(\frac{P}{\epsilon} \right) \tag{2}$$

where

L = support span,
a = loading span,
b = width of specimen,
t = thickness of specimen,
P = total load on specimen,
ϵ = midspan strain, and
P/ϵ = slope of load-strain diagram.

Table 2 shows the results of all static tests in terms of the modulus of elasticity and the strain at the proportional limit. Proportional limit data are of interest in the study of amplitude effects. As shown in Table 2, some of the materials were obtained by the authors directly from PPG Industries (PPG)

TABLE 2—Static flexure test data.[a]

Material	$\bar{E}$ (10^6 psi)	$\bar{E}$ (GPa)	S (10^6 psi)	S (GPa)	$S/\bar{E}$, %	$\bar{\epsilon}_p$, 10^{-3}	σ, 10^{-3}	$\sigma/\bar{\epsilon}_p$, %	T, °C	H, %
PPG SMC-R25[b]	2.17	(14.96)	0.0529	(0.365)	2.43	2.68	1.10	41.1	19	...
PPG SMC-R65[b]	2.40	(16.55)	0.240	(1.655)	10.0	2.27	0.125	5.50	19	...
PPG SMC-R25 matrix resin[b]	1.32	(9.10)	0.0253	(0.174)	1.92	2.33	1.14	48.9	24	40
PPG SMC-R65 matrix resin[b]	0.514	(3.54)	...	...	...	7.25	...	...	24	44
PPG SMC-R65[c]	2.71	(18.68)	0.107	(0.738)	3.94	1.83	0.263	14.3	19	35
PPG XMC-3(L)[c]	5.47	(37.72)	0.0909	(0.627)	1.66	2.02	0.0625	3.09	19	35
PPG XMC-3(T)[c]	2.06	(14.2)	0.0255	(0.176)	1.24	1.00	0.100	10.0	10	35
OCF SMC-R25[c]	1.23	(8.48)	0.0922	(0.636)	7.48	4.80	0.622	13.0	22	...
OCF C20/R30(L)[c]	4.09	(28.2)	0.385	(2.655)	9.40	2.20	0.163	7.42	22	...
OCF C20/R30(T)[c]	1.39	(9.58)	0.242	(1.669)	17.4	1.42	0.908	64.1	20	33

[a] $\bar{E}$ = average value of modulus of elasticity from three specimens, $\bar{\epsilon}_p$ = average value of strain at proportional limit, S = standard deviation on E, σ = standard deviation on ϵ_p, T = temperature, and H = relative humidity.
[b] Obtained directly from PPG Industries.
[c] Supplied by GM.

and some were supplied indirectly by PPG and Owens-Corning Fiberglas (OCF) through General Motors (GM). All sheet molding compound (SMC) materials and matrix resin materials were assumed to be isotropic, so specimens were cut in one direction only. The hybrid continuous/chopped fiber composites (PPG XMC-3 and OCF C20/R30) are known to be highly anisotropic, so specimens were cut along both longitudinal and transverse axes. It appears that the PPG SMC-R25 is nearly twice as stiff as the OCF SMC-R25. As shown later, dynamic test results show the same pattern. Since both materials have the same amounts of glass fibers (25 percent by weight), the difference must be in the matrix resins. As expected, the OCF C20/R30 transverse modulus is about the same as the modulus for the OCF SMC-R25, since the chopped fiber reinforcement governs the transverse behavior of the C20/R30. The PPG SMC-R25 matrix resin is more than twice as stiff as the PPG SMC-R65 resin; this is apparently due to differences in fillers. There is also a slight difference between the PPG SMC-R65 obtained directly from PPG and that supplied by GM. The expected anisotropic behavior of the hybrid composites is reflected in the differences between longitudinal and transverse stiffness.

Forced Vibration Technique

A detailed description of the experimental technique and apparatus has been given in Ref 4. The double cantilever specimen and associated transducers are shown in Fig. 1. Since the development of the original apparatus, we have designed a new apparatus that uses an aluminum clamshell vacuum/environmental chamber to house the specimen. The chamber is mounted on the body of an MB model C11 shaker. The specimen mounting pedestal is attached to the shaker armature, and a flexible diaphragm provides a dynamic seal between the chamber and the armature. The chamber is evacuated to less than 10-mm mercury pressure with a mechanical vacuum pump in order to minimize air damping at low frequencies and corresponding large amplitudes. Air damping can completely mask out material damping if amplitudes are large enough [4].

Calibration of the new apparatus was accomplished by testing a 2024-T351 aluminum alloy specimen as in Ref 4. By comparing measured damping with predicted thermoelastic damping (which is the predominant mechanism for most structural metals in flexural vibration), the capabilities of the apparatus can be determined.

The SMC specimen configuration is shown in Fig. 2. Specimens were cut from standard 3.17-mm (0.125-in.)-thick precured plaques. Polyester mounting shoulders were formed by pouring the uncured resin into removable Teflon molds and allowing it to set up. Most specimens were made by joining two pieces together at the shoulder. This was necessary to obtain low frequency data from standard plaques. This is discussed later under

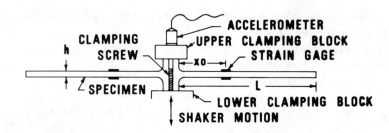

FIG. 1—*Double cantilever beam specimen.*

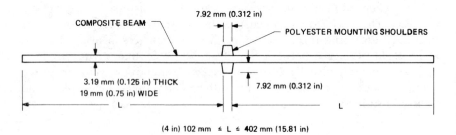

FIG. 2—*Composite specimen configuration.*

Experimental Results. Testing began at the longest specimen lengths; data were obtained up through the fifth mode at each length before cutting off to shorter lengths. In this manner, the nominal frequency range of 10 to 1000 Hz was covered. Maximum strain amplitudes ranged up to 0.0017.

Calculation of Complex Modulus

The real and imaginary parts of the complex Young's modulus were calculated as follows. The storage modulus E' was found by using the frequency equation for the cantilever beam specimen:

$$\omega_r = \frac{(\lambda_r L)^2}{L^2}\left[\frac{E'I}{\rho_m A}\right]^{1/2} \tag{3a}$$

or

$$E' = \frac{\omega_r^2 L^4 \rho_m A}{(\lambda_r L)^4 I} \tag{3b}$$

where

A = cross-sectional area of specimen,
I = area moment of inertia of specimen cross section,

L = specimen length,
ρ_m = specimen density,
λ_r = eigenvalue of rth mode, and
ω_r = angular resonant frequency of rth mode.

As shown in Ref 4, the loss factor η can be defined as

$$\eta = \frac{D_s}{2\pi U_s} \qquad (4)$$

where

D_s = energy dissipated per cycle, and
U_s = strain energy at maximum displacement.

For the cantilever specimen in steady-state resonant vibration, it is shown that

$$\eta = \frac{\lambda_r \phi_r'''(0)\phi_r''(x_0)h\ddot{a}_0}{2L\omega_r^2 \cdot \epsilon(x_0)} \qquad (5)$$

where

$$h = \text{specimen thickness,}$$
$$\ddot{a}_0 = \text{base acceleration amplitude,}$$
$$\phi_r(x), \phi_r''(x), \phi_r'''(x) = \text{mode shape function and derivatives,}$$
$$x = \text{distance along specimen from base,}$$
$$x_0 = \text{location of strain gage, and}$$
$$\epsilon(x) = \text{bending strain at location } x.$$

Once the storage modulus and the loss factor are found from Eqs 3 and 5, respectively, the loss modulus E'', is found from $E'' = \eta E'$. The maximum strain amplitude at the base of the specimen is related to the measured strain amplitude by

$$\epsilon_{\max} = \epsilon(0) = \epsilon(x_0)\frac{\phi_r''(0)}{\phi_r''(x_0)} \qquad (6)$$

Experimental Results

Figure 3 shows the results of several tests of the aluminum calibration specimen. The intent of these tests is to estimate the accuracy of damping measurements by using a specimen for which the damping can be calculated. The Zener thermoelastic theory is quite accurate in predicting damping for structural metals under flexural vibration [4,5,7]. The subsequent experiments with SMC showed that the actual material damping in SMC is

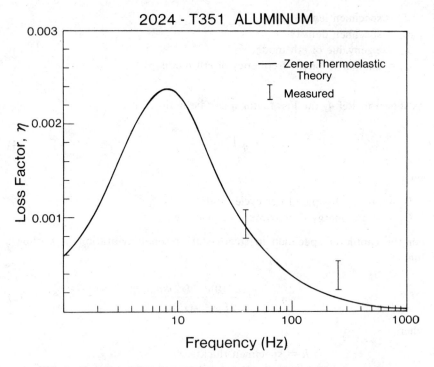

FIG. 3—*Variation of damping with frequency for aluminum calibration specimen.*

several orders of magnitude higher than the thermoelastic prediction, however. The thermoelastic loss factor η_T is

$$\eta_T = \frac{\alpha^2 E T}{C} \cdot \frac{\omega\tau}{1 + \omega^2\tau^2} \tag{7}$$

and the relaxation time is

$$\tau = \frac{h^2 C}{\pi^2 K} \tag{8}$$

where

    α = coefficient of thermal expansion,
    E = modulus of elasticity,
    T = absolute temperature,
    ω = angular frequency of vibration,
    C = specific heat per unit volume, and
    K = thermal conductivity.

The required thermal properties are shown in Table 3. In physical terms, when a beam vibrates in flexure, the fibers in tension are cooled, while those in compression are heated. The resulting temperature gradient across the beam causes a continual flow of heat back and forth across the beam and this heat flow involves the loss energy. At high frequencies ($\omega \gg 1/\tau$), the heat has no time to flow across the beam, so that the process may be assumed to be adiabatic. At low frequencies ($\omega \ll 1/\tau$), the temperature gradient is so small that there is very little heat flow. However, if the period of vibration is about the same as the relaxation time for temperature equalization ($\omega \simeq 1/\tau$), there is a condition of maximum energy dissipation.

The two scatter bands in Fig. 3 represent the range of measured loss factors in a beam with $L = 254$ mm (10 in.), $h = 3.17$ mm (0.125 in.), and width $= 19$ mm (0.75 in.), in first and second mode vibration. This shows that the apparatus gives good damping data for loss factors greater than about 0.0005. Below this, it appears that background damping in the apparatus begins to mask out material damping. Fortunately, the SMC loss factors were on the order of 0.01, well within the capabilities of the apparatus.

Figure 4 shows the effect of air damping of a composite specimen at large amplitudes. The loss factors in air and in vacuum are plotted versus the amplitude-to-thickness ratio because the theoretical loss factor due to aerodynamic drag is proportional to this ratio [4]. Figure 4 shows that, for sufficiently small amplitudes, the air damping is negligible. This has important implications for future environmental tests. Since the environmental tests cannot be conducted under vacuum conditions, the amplitudes must be kept low enough so that air damping does not significantly affect the results. One might ask: "If air damping is so large, why worry about material damping?" The answer is that, due to the nature of the cantilever specimen, the vibration amplitudes are much greater than those found in actual automobile or other structures. Thus air damping is usually negligible in the actual structure, but not in the cantilever specimen.

Figure 5 shows that the loss factor for PPG SMC-R25 is essentially independent of amplitude up to the maximum strain amplitude of about

TABLE 3—*Physical properties used in calculation of thermoelastic damping.*

Parameter	2024 Aluminum[a]	PPG SMC-R25[b]
Coefficient of linear expansion, α, ($10^{-6}/°C$)	23.2	28.8
Specific heat, C, (cal/g per °C)	0.21	0.255
Modulus of elasticity, E, (GPa)	73.1	14.96
Thermal conductivity, k, (N/s per °C)	126.0	0.138

[a] After Ref 7.

[b] Data from personal communication with C. Luther, PPG Industries, Pittsburgh, Pa., Feb. 1979.

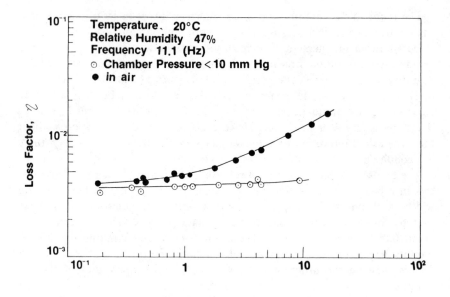

FIG. 4—*Variation of damping with amplitude for OCF C20/R30 specimen.*

0.0005. Similar behavior was observed in tests of the other composites. Thus the complex modulus notation may be used for all materials tested here. More recent tests have gone up as high as 0.0017 maximum strain with no evidence of amplitude dependence. Previous experiments with 3M Scotchply continuous fiber glass/epoxy composites [6] showed that damping is quite sensitive to the effects of fatigue and microstructural damage. Static flexure tests of the SMC materials showed that stiffness is independent of amplitude up to the proportional limit (Table 2); thus no microstructural damage is expected at these amplitudes.

Figures 6 to 9 show typical measured storage moduli, loss moduli, and loss factors for PPG SMC-R65 and PPG SMC-R25 at various frequencies in the nominal range 10 to 1000 Hz. These data were generated by exciting the first and second modes at various lengths, while the data in Figs. 10 to 12 were generated up through the fifth mode. Testing at higher modes meant that the specimen did not have to be cut off as many times to cover the desired frequency range. The PPG SMC-R25 and the PPG SMC-R65 were the only materials for which more than one specimen was tested. In Figs. 6 to 9, Specimen A refers to a one-piece specimen, whereas Specimen B refers to a two-piece specimen with a joint at the clamping shoulder. The two-piece specimens were necessary to get low-frequency data from the standard plaques. As shown in Figs. 6 to 9, the agreement between the properties of

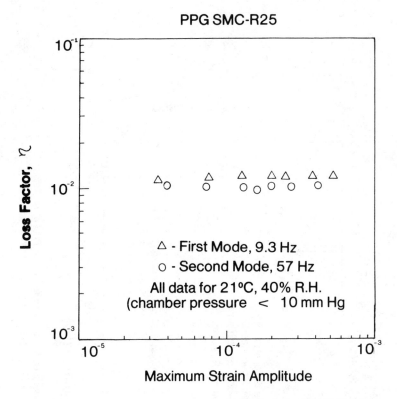

FIG. 5—*Loss factor versus maximum strain amplitude for PPG SMC-R25.*

the one-piece and two-piece specimens is good. Once the validity of the two-piece specimen data was established, all the remaining data shown in Figs. 10 to 12 were generated on two-piece specimens. The overlap between the data for the different modes and the good agreement with static flexure data supports the validity of the dynamic data. Most of the scatter in the loss factor–loss modulus data is believed to be caused by (*1*) proximity of strain gages to nodal points for some of the higher modes, (*2*) imbalance of specimens for higher modes (only the first mode was balanced), and (*3*) variations in temperature and humidity between tests. The measured storage modulus is only slightly affected by the amount and location of balancing weights and by the variations in environmental conditions, and is not affected at all by errors in strain measurement. With this in mind, it appears that all the properties are practically independent of frequency within the test range. The fact that the apparent stiffness does not decrease at higher frequencies also indicates that shear effects are not significant.

Using Eqs 7 and 8 with the PPG SMC-R25 properties from Table 3, it was

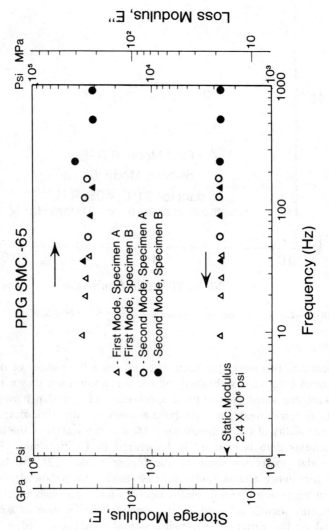

FIG. 6—*Storage modulus and loss modulus versus frequency for PPG SMC-R65.*

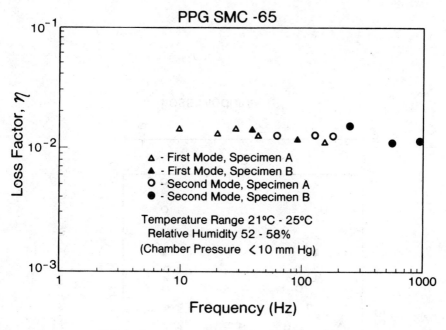

FIG. 7—*Loss factor versus frequency for PPG SMC-R65.*

estimated that the thermoelastic loss factor η_T was less than 10^{-5} over the frequency range 10 to 1000 Hz. Measured SMC-R25 loss factors are greater than 0.01, however (Fig. 9). Thermoelastic dissipation in SMC materials is lower than for aluminum primarily because of the differences in stiffness and thermal conductivity. Thus the measured damping in SMC must be caused by other effects, such as viscoelastic matrix behavior and interfacial dissipation. The loss factors for all composites tested were easily an order of magnitude greater than the aluminum loss factors over the range 10 to 1000 Hz. In addition, the aluminum has very low damping at higher frequencies, whereas the composites had essentially constant damping over the entire frequency range.

A comparison of the complex moduli of all the materials tested is given in Figs. 10 to 12. A general observation here is that the materials having the greatest storage moduli have the lowest loss factors, and vice versa. This is because the behavior of the low stiffness materials [the matrix resins, SMC-R25, XMC-3(T), and C20/R30/T)] is matrix-controlled, whereas the behavior of the high stiffness materials [XMC-3(L), C20/R30(L)] is fiber-controlled. Since most of the damping occurs in the matrix resin, matrix-controlled deformation should produce more dissipation than fiber-controlled deformation. However, it is interesting to note that, although the

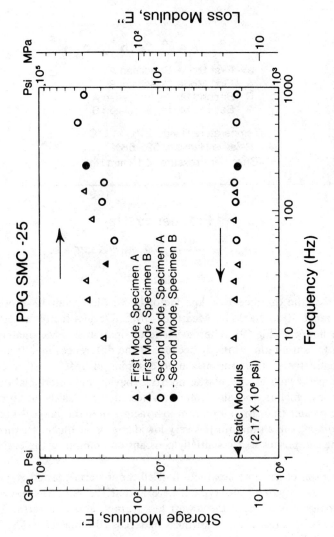

FIG. 8—*Storage modulus and loss modulus versus frequency for PPG SMC-R25.*

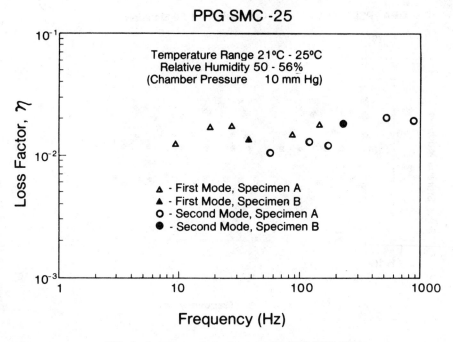

FIG. 9—*Loss factor versus frequency for PPG SMC-R25.*

stiffness of the PPG SMC-R25 is nearly twice that of the OCF-R25, the loss factors are practically the same. Similarly, the stiffness of the PPG SMC-R25 matrix resin is more than twice that of the PPG SMC-R65 matrix resin, but the loss factors are about the same. Although the loss factors of all materials range between 0.004 and 0.027, most of the data fall between 0.01 and 0.02.

Conclusions

1. The complex moduli of all test materials are practically independent of amplitude and frequency within the ranges tested.

2. The materials having greatest stiffness [the XMC-3(L) and C20/R30(L)] generally have the lowest damping, while the materials having lowest stiffness [the matrix resins and C20/R30(T)] generally have the greatest damping. This is because the behavior of the stiffest materials is fiber-controlled, but most of the damping occurs in the matrix resin.

3. Measured damping in the SMC materials is several orders of magnitude greater than the estimated thermoelastic damping. This means that the measured damping is caused by other effects, such as viscoelastic matrix behavior or interaction between matrix and fibers.

4. Measured SMC damping is at least one order of magnitude greater

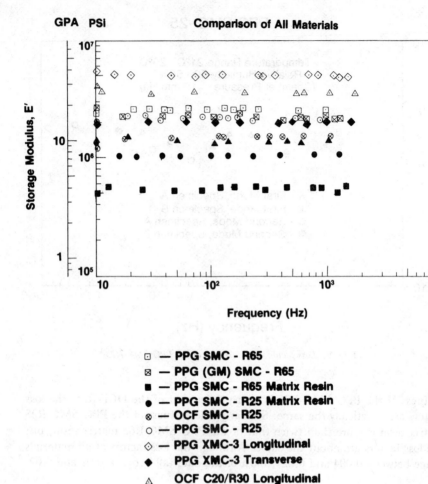

FIG. 10—*Comparison of all materials: storage moduli.*

than the damping in 2024-T351 aluminum of the same thickness over the test frequency range.

5. The PPG SMC-R25 has nearly twice the stiffness of the OCF SMC-R25, but the loss factors are about the same.

6. The PPG SMC-R25 matrix resin has more than twice the stiffness of the PPG SMC-R65 resin, probably because the former contains a filler. The loss factors of the two resins are about the same, however.

7. Measured complex moduli of the test materials show good agreement

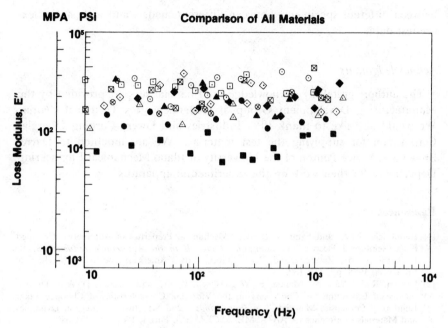

FIG. 11—*Comparison of all materials: loss moduli. See Fig. 10 for legend.*

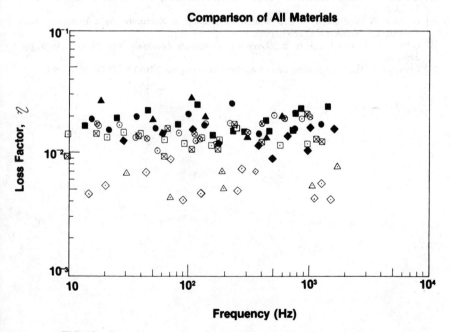

FIG. 12—*Comparison of all materials: loss factors. See Fig. 10 for legend.*

between different specimens, between different modes, and with static flexure test data.

Acknowledgments

The authors gratefully acknowledge the financial support provided by the Manufacturing Development Group at General Motors Technical Center. We would also like to thank PPG Industries and Owens-Corning Fiberglas Corporation for supplying the test materials. We are indebted to Darrel Brown and Vance Penton of the University of Idaho Mechanical Engineering Department for their work on the experimental apparatus.

References

[1] Heimbuch, R. A. and Sanders, B. A., "Mechanical Properties of Automotive Chopped Fiber Reinforced Plastics," in *Composite Materials in the Automobile Industry,* S. V. Kulkarni, C. H. Zweben, and R. B. Pipes, Eds., American Society of Mechanical Engineers, 1978, pp. 111–139.
[2] Gibson, R. F., Yau, A., Mende, E. W., Osborn, W. E., and Riegner, D. A., "The Influence of Environmental Conditions on the Vibration Characteristics of Chopped Fiber Reinforced Composite Materials," in *Proceedings,* 22nd Structures, Structural Dynamics and Materials Conference (AIAA/ASME/ASCE/AHS), Part 1, 1981, pp. 333–340.
[3] Lazan, B. J., *Damping of Materials and Members in Structural Mechanics,* Pergamon Press, Oxford, 1968.
[4] Gibson, R. F. and Plunkett, R., *Experimental Mechanics,* Vol. 11, No. 8, Aug. 1977, pp. 297–302.
[5] Granick, N. and Stern, J. E., "Material Damping of Aluminum by a Resonant Dwell Technique," NASA TN-D-2893, 1965.
[6] Gibson, R. F. and Plunkett, R., *Journal of Composite Materials,* Vol. 10, Oct. 1976, pp. 325–341.
[7] Crandall, S. H., "On Scaling Laws for Material Damping," NASA TN D-1467, 1962.

S. S. Wang[1] and T. P. Yu[1]

Statistical Fracture Initiation in Randomly Oriented Chopped-Mat Fiber Composites Subjected to Biaxial Thermomechanical Loading

REFERENCE: Wang, S. S. and Yu, T. P., **"Statistical Fracture Initiation in Randomly Oriented Chopped-Mat Fiber Composites Subjected to Biaxial Thermomechanical Loading,"** *Short Fiber Reinforced Composite Materials, ASTM STP 772*, B. A. Sanders, Ed., American Society for Testing and Materials, 1982, pp. 151-166.

ABSTRACT: Owing to the inherently heterogeneous microstructure of constituent elements and the random orientation of fiber mats, initiation of local fracture in chopped-mat fiber composite materials is statistical in nature. An analytical approach is presented in this paper for studying the fracture initiation. The study is formulated on the basis of the statistical failure theory and fundamental concept of mechanics of fracture for brittle fiber composites. The heterogeneity of chopped fiber mats with random orientations and the effect of biaxial thermomechanical stress are incorporated into the Weibull strength theory. A computer simulation procedure with the aid of a numerical stress analysis is introduced. Estimation of the location of fracture initiation in a statistical sense is conducted for the disk-type chopped-mat fiber composite. Selected results are presented to illustrate the fundamental nature of the statistical fracture initiation behavior in this class of composite material.

KEY WORDS: chopped fiber mat, fiber reinforced composite, fracture initiation, statistical strength theory, thermomechanical stresses, failure probability, Weibull's weakest-link model

Recent advances in fiber and polymer science and technology have led to the use of short-fiber reinforced composite materials in many engineering structures and components such as bulk molding compounds (BMC) and sheet molding compounds (SMC) for automobiles [1–4].[2] Because of the unique combination of desired thermal and mechanical properties and pro-

---

[1] Associate Professor and Graduate Research Assistant, respectively, Department of Theoretical and Applied Mechanics, University of Illinois, Urbana, Ill. 61801.
[2] The italic numbers in brackets refer to the list of references appended to this paper.

cessibility of the material, short-fiber composites such as chopped-mat glass fiber reinforced epoxy are currently considered in the construction of pressure vessels for transport and storage of cryogenic fluids (for example, liquid natural gas, liquid hydrogen, etc.). Many fundamental problems associated with this relatively new class of composites need to be studied before the full potential of the material can be achieved. One major concern is structural and functional reliability in a fail-safe design of the composite materials and structures. The structural reliability is concerned with the loss of load-bearing capacity due to fracture during service. The functional reliability refers to failure to achieve the desired performance other than structural requirements (for example, leakage of fluid through the wall of a pressure vessel in the present case). The leakage problem has been a major consideration in pressure vessel design, and the study of the leak-before-fracture concept has provided useful information in the fracture control technology of homogeneous metals [5,6]. However, research progress has been relatively slow for the problem in heterogeneous fiber composites, especially the short-fiber composite system. The importance and consequence of the functional and structural failure have been well recognized, and are observed to be related directly to the initiation and subsequent growth of cracks.

The problem of initiation and growth of cracks in a short-fiber composite is extremely complicated, involving not only the heterogeneity of fibers and resin matrix but also the inherently large statistical scatter of strength properties. The apparently random distribution of reinforcing chopped-mat fibers introduces an additional difficulty in the investigation. Under a given thermomechanical loading condition, the detailed nature of fracture initiation and the extension of damage in the short-fiber composite are difficult to predict based on classical deterministic fracture mechanics for homogeneous solids. Even an approximate solution for the problem simply does not exist to the authors' knowledge. This necessitates a fundamental study of the problem. In this paper, an analytical method is presented for studying fracture initiation in randomly oriented, chopped-mat fiber composites subjected to general thermomechanical loading.

The basic notion underlying this study is that, in the highly heterogeneous composite, fracture initiation generally results from local failure of the material due to the combined effect of the worst flaw and the most unfavorable strain in the solid. This paper attempts to explore this concept by developing an analytical approach to study the fracture behavior from a statistical point of view and to elucidate the implications from the results obtained. Some experimental observations of initiation of fracture in biaxially loaded, chopped-mat composites are discussed first. Modeling and basic formulation of the problem are given, and a unique parameter that gives a measure of probabilistic fracture initiation in the composite is introduced. A procedure for computer-aided numerical simulation of the phenomenon is then presented. Results are shown for composite disks used in typical thermal shock experiments.

Fracture probability and initiation sites and the influence of biaxial thermomechanical stress are examined for composite systems with various Weibull parameters. This study is one of the first to incorporate the random microstructure of chopped fiber-mat composites into a fracture analysis and to include the
effect of biaxial thermomechanical stress in the Weibull statistical failure
theory for the short fiber composites. The method of approach is of practical
use in the reliability design and testing of randomly oriented chopped-mat
fiber reinforced composites. The results advance our current understanding
of the behavior of statistical fracture initiation in this class of material.

Experimental Observations

In order to investigate the failure behavior of chopped-mat fiber reinforced
composites subjected to severe thermomechanical loading, a laboratory test
was developed [7] by employing a composite disk specimen with a tapered
cross section supported around its perimeter and subjected to lateral loading
in a cryogenic environment (Fig. 1). The objective of the experiments was to
simulate, as closely as possible, the combined thermal, mechanical, and environmental loading—that is, the so-called thermal shock—that the composite would encounter in service as a component of a pressure vessel containing
cryogenic fluids [8]. Cryogenic strain gages were mounted on the surfaces of
the specimen, and the distribution of strain was measured in the test. The
maximum strain always occurred approximately at the center of the panel.
Metallographic examination was conducted for a number of test specimens
after leakage experiments. Fracture initiation was generally deduced from
the morphology of surface cracking in the experiment. Locations of the fracture initiation were observed to occur seldomly at the center of a composite
disk [8]. This may be a statistical effect resulting from the combination of the
worst flaw and strain at the critical location. Similar fracture phenomena
were reported to occur in disk specimens of different grades of beryllium,
"mechanite" cast iron, pyrophyllite, and Pyrex glass under biaxial tensile
stresses [9]. For homogeneous, brittle materials, fracture initiation at a
distance away from the center is a simple manifestation of size effects. However, for the fiber composite studied here, the chopped fiber-mat reinforcements may play a dominant role in controlling the fracture. Thus it is necessary to include the microstructural aspects of the material in studying the
statistical fracture initiation problem.

Modeling

In order to study the problem more rigorously, an analytical approach is
proposed. The chopped-mat fiber composite disk is modelled as an assem-
blage of composite elements (Fig. 2). Within each element, the fiber mat has
its principal material axes oriented at an angle ϕ with respect to structural

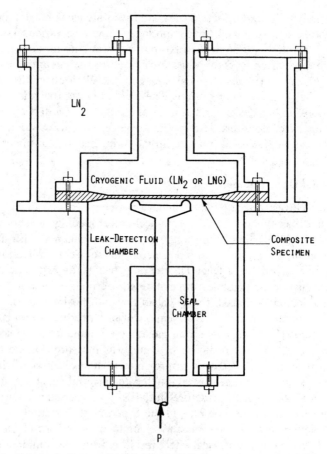

FIG. 1—*Schematic diagram of biaxial thermal shock test of composite disk specimen.*

coordinates. It is noted that fiber orientations in the composite elements are random, and may be expressed by

$$\phi = Dx \tag{1}$$

where x is a random number generated by a computer-aided random number generator D, and ϕ has a value ranging from 0 to π. Each individual element has a fiber-volume fraction, V_f, consistent with that of the overall composite panel, and a dimension, $(r\Delta\theta\Delta r)$, of the size of a chopped fiber mat in the composite, where r, Δr, and $\Delta\theta$ are as defined in Fig. 2. The element is assumed to have orthotropic elastic properties in local material coordinates. For the convenience of later analytical development, the chopped-mat composite elements are placed in a concentric annular arrangement. Material constants of each element with respect to the structural axes are related to

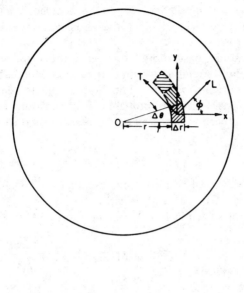

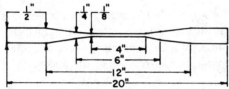

FIG. 2—*Modeling of chopped-mat fiber composite and geometric configuration of test specimen.*

the element fiber orientation ϕ in Eq 1 through a transformation. The composite disk is subjected to both mechanical and thermal loading (Fig. 1).

Basic Formulation

During the last decade a significant amount of research has been directed to assess the reliability of composite materials and structures. Various statistical theories have been proposed to account for the inherently large scatter in the failure strength [10-15]. The three-parameter Weibull failure theory, based on the weakest-link model, has been shown to be a useful and versatile means of describing the statistical strength of a composite material [16,17]. Though arguments have arisen concerning the theoretical basis for using this distribution function, it provides a satisfactory phenomenological fit of failure data for many composite systems. Other statistical models have also been proposed for estimation of the failure behavior of fiber composite materials such as the parallel model [18], the dispersed fracture model [19], and the cumulative weakening model [20], etc. A detailed discussion of the advantages and disadvantages of using the various composite failure models has been given in Ref 21.

In the current study, basic formulation lies in the introduction of the commonly used Weibull strength distribution function for each chopped-mat fiber composite element. Fracture initiation in the test specimen is then evaluated on the basis of failure probabilities of constituent elements by using a convenient procedure that will be discussed subsequently.

Statistical Fracture of a Chopped-Mat Fiber Composite Element

Based on the weakest-link model, the cumulative probability distribution function of failure, $G(s)$, at a given stress level s for a chopped-mat fiber composite element, shown in Fig. 2, may be expressed as

$$G(s) = 1 - \exp\left\{-\int_{V*}\left[\frac{\sigma(s) - \sigma_u}{\sigma_0}\right]^m dV\right\} \tag{2}$$

where σ_0, σ_u, and m are Weibull's scaling, location, and shape parameters, respectively, and $V*$ is the volume of a unit composite element under a general stress state $\sigma(s)$. In a uniaxial tension test, the reference stress may be taken as the nominal applied stress σ_∞ of the specimen. Then Eq 2 can be written as

$$G(\sigma_\infty) = 1 - \exp\left[-V\left(\frac{\sigma_\infty - \sigma_u}{\sigma_0}\right)^m\right] \tag{3}$$

Multiaxial State of Thermomechanical Stress

In the present problem, each chopped-mat fiber composite element is subjected to biaxial thermomechanical stress, which varies significantly with the location in the composite, mainly due to the random distribution of fiber mat orientations. To apply the Weibull form of statistical fracture characterization to a situation other than simple tension, the problem associated with the response of a flaw to a multiaxial state of stress must be examined. Incorporation of this complication into Eq 2 is necessary.

Since fiber orientations in composite elements are randomly distributed, and since crack-like flaws (Fig. 3) are observed to orient along the fiber direction in each element, the current Weibull failure theory of Eq 2 for a composite element involving biaxial stresses may be modified conveniently. Corten [22] suggested that the flaws considered in this case would extend only in a Mode I fracture when the combination of flaw length a and tensile

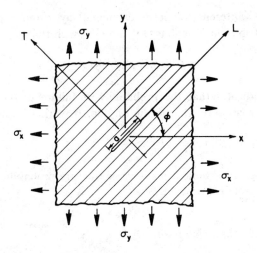

FIG. 3—*Flaw geometry and structural and material coordinates in a chopped-mat fiber composite element.*

stress normal to the flaw (that is, the transverse stress) σ_t satisfies a relation of the form

$$\sigma_t \sqrt{a} = \text{constant} \tag{4}$$

It can be shown easily that, for a biaxially stressed composite element with a random fiber orientation, the probability of fracture for the element may be expressed as

$$G(\sigma_t) = 1 - \exp\left[-V^*\left(\frac{\sigma_t - \sigma_u}{\sigma_0}\right)^m \right] \tag{5}$$

Probabilistic Fracture Parameter of a Composite Element

Assuming Weibull's location parameter $\sigma_u = 0$, and taking logarithmic operations on both sides of Eq 5, one can introduce a new parameter P_f as

$$P_f \equiv m \log \sigma_t + \log V^* + C \tag{6a}$$

where P_f is related to $G(\sigma_t)$ by

$$P_f = \log \{ -\log [1 - G(\sigma_t)] \} \tag{6b}$$

and $C = -m \log \sigma_0$.

Consider two different composite elements, say, i and j, with cumulative probability distribution functions $G^{(i)}$ and $G^{(j)}$ such that

$$G^{(i)}(\sigma_t) \geq G^{(j)}(\sigma_t) \tag{7}$$

This simply indicates that the probability of failure in the ith element is greater than, or equal to, that in the jth one. It further implies that

$$P_f^{(i)} \geq P_f^{(j)} \tag{8}$$

Thus P_f provides a convenient measure of fracture probability of a chopped-mat fiber composite element.

Fracture Initiation in a Composite Disk

Consider the composite disk shown in Fig. 2. The chopped-mat composite elements are subjected to biaxial thermomechanical stresses that are to be calculated by a solution scheme given in the next section. The element has a volume $V^* = r\Delta\theta\Delta rt$, which may be chosen as the size of the chopped fiber mat introduced during manufacturing. The fracture parameter P_f can be evaluated for each element by

$$P_f^{(i)} = m \log \sigma_t^{(i)} + \log(r^{(i)}\Delta\theta^{(i)}) + \log \Delta r^{(i)} + C^* \tag{9}$$

where C^* is a constant defined as $(C + \log t)$. It is noted that $P_f^{(i)}$ is a function of the element position, $r^{(i)}$ and $\Delta\theta^{(i)}$, in the composite disk. For simplicity without loss of generality, the fracture parameter for an annulus at a distance r from the center consisting of n randomly oriented chopped-mat composite elements with a uniform size may be evaluated by

$$E[P_f^{(i)}] = m E[\log \sigma_t^{(i)}] + \log r + \log(\Delta r^{(i)}\Delta\theta^{(i)}) + C^* \tag{10}$$

where $E[\ ]$ is the mathematical mean of the associated quantities in the brackets, defined as

$$E[P_f^{(i)}] = \frac{1}{n} \sum_{i=1}^{n} P_f^{(i)} \tag{11}$$

More explicitly, $E[P_f^{(i)}]$ in Eq 10 is a function of r also and represents the mean of failure probabilities of all elements in the annulus located at distance r from the center. Comparing the values of $E[P_f^{(i)}]$ of all annuli in a disk specimen provides the information on fracture initiation in the composite.

Simulation Procedure of Statistical Fracture Initiation in the Composite

Just as many complex field problems are amenable to numerical techniques, the complicated fracture initiation problem in short fiber composites can also be studied by a computer-aided simulation method in conjunction with the aforementioned probabilistic failure theory. By setting up a model like the one given in the previous section, one can conveniently analyze the response of the whole system when a suitable random number generator and a proper solution scheme are introduced. Thus main ingredients of the simulation procedure consist of three major parts: random number generation, local stress calculation, and appropriate probabilistic failure criteria for the elements and the composite disk.

Random Number Generation

As shown in Fig. 2, n chopped-mat fiber composite elements are partitioned in an annulus of the composite disk. To simulate random fiber orientations in the elements, n random numbers within the interval $[0, \pi]$ need to be generated. By using a digital computer, generation of a set of random numbers with a uniform distribution is a standard practice in mathematical statistics. A built-in library subroutine [23] in the IBM 360 computer system of the University of Illinois was used in the present study to generate a sequence of uniformly distributed, random floating numbers in the interval desired. (The numerical technique is based on the so-called multiplicative congruential method; details of the method are reported in Ref 24.) Typical results for several cases are given in Fig. 4 for illustrative purposes. The histograms in the figure show frequencies of fiber orientations of 150 elements in an annulus of the composite disk. The population of random fiber orientations generated approximates closely to a uniform distribution over $[0, \pi]$ in each case.

Stress Solution for a Composite Mat Element

It is clear from Eq 9 that determination of probabilistic fracture requires information of the state of thermomechanical stress in each of the composite elements. The approximate, continuous mechanical and thermal strain fields, ε_m and ε_T, are evaluated first by a solution scheme employing a conventional finite-element method for the composite subjected to a prescribed loading condition shown in Fig. 1. By using axisymmetric, solid-ring elements for the strain calculation, a very small opening exists at $r = 0$ in the finite-element mesh arrangement. Corresponding thermomechanical stresses in each individual element within an annulus are then evaluated by

$$\underline{\sigma}^{(i)} = \underline{Q}^{(i)} (\varepsilon_m + \varepsilon_T) \qquad (12)$$

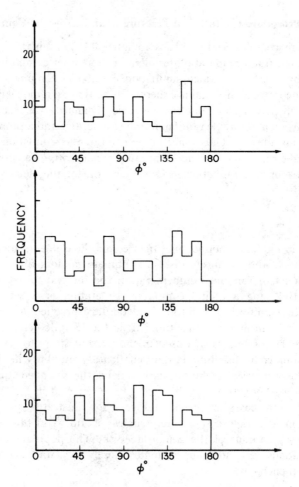

FIG. 4—*Histograms of random number generation for fiber orientations in chopped-mat composite elements.*

where $Q^{(i)}$ is the stiffness matrix of the ith composite element with fiber orientation $\phi^{(i)}$.

The stresses $\bar{\sigma}^{(i)}$ associated with principal material coordinates of the element are determined by a transformation $T(\phi^{(i)})$:

$$\bar{\sigma}^{(i)} = T(\phi^{(i)}) \, \sigma^{(i)} \tag{13}$$

where

$$\bar{\sigma}^{(i)} = \{\sigma_\ell^{(i)}, \sigma_t^{(i)}, \sigma_{\ell t}^{(i)}\} \tag{14}$$

with ℓ and t referring to the fiber and transverse directions, respectively.

Fracture Initiation Criterion

The study of statistical fracture initiation in the chopped-mat fiber composite model requires a direct evaluation of the fracture parameter $P_f^{(i)}$ of each composite element. By using the aforementioned random number generation scheme to simulate random fiber orientations, and using the finite-element method for calculating local stresses, one can easily determine the values of $P_f^{(i)}$ for all composite elements in the disk specimen. Thus a plot of $E[P_f^{(i)}]$ of each annulus versus r provides the variation of failure probability with radial distance in composite systems having different Weibull parameters, m and σ_0. The position r_f^* corresponding to the maximum value of $E[P_f^{(i)}]$ identifies the radial distance in the composite disk, where fracture initiation is expected to occur. A better assessment of the statistical fracture characteristic in the chopped-mat fiber composite requires a full-scale, computer-aided Monte Carlo simulation. The numerical outcome provides a description of the fracture phenomenon for the composite, which is observed in laboratory experiments [7,8].

Numerical Examples and Discussion

To illustrate the present method of approach, the thermally shocked composite disk discussed in the section on experiments has been investigated. Consider the composite disk subjected to both thermal and mechanical loading as shown in Fig. 1. The composite consists of chopped-strand, E-glass fiber mats and epoxy resin matrix with the following material properties:

$$E_f = 68.9\,\text{GPa}\,(10 \times 10^6\,\text{psi}) \qquad E_m = 3.45\,\text{GPa}\,(0.5 \times 10^6\,\text{psi})$$

$$\nu_f = 0.21 \qquad \nu_m = 0.35$$

$$\alpha_f = 0.5 \times 10^{-5}/°\text{C} \qquad \alpha_m = 6 \times 10^{-5}/°\text{C}$$

$$\Delta T = -210°\text{C}$$

$$V_f = 0.55$$

Following the aforementioned solution procedure, strains at any location in the composite disk are evaluated first. Detailed numerical results have been reported elsewhere [7] and are not repeated here. It is only stated that the maximum strain almost always occurs at the center of the composite disk. However, as mentioned in the experiments, locations of fracture initiation, as evidenced in many pressurized leakage tests [8], are seldomly observed to occur at the center of the specimens. Within the gage section, which is the region of major interest, 150 randomly oriented, mat-composite elements are used to model a typical annulus in the composite panel. It is believed that the population of random fiber orientations generated is enough to represent the common statistical nature of the fiber-mat microstructure in the material. By evoking the constitutive equations of the composite elements, thermomechanical stresses at any location are determined. The failure parameter of each composite element is then evaluated by the procedure suggested in previous sections. The value of $E[P_f]$, normalized by its value at $r \sim 0$, is a function of the radial coordinate r. The maximum $E[P_f]$ and its associated radial distance r (defined as r_f^* hereafter as shown in Fig. 5) provide fundamental information on fracture initiation in the composite. It is noted that, in computing $E[P_f]$, $\Delta r^{(i)}$ and $\Delta \theta^{(i)}$ are kept constant for all elements in an annulus. Thus, even though $E[\log \sigma_t^{(i)}]$ is found to decrease with the increase of r in all cases studied, failure probability $E[P_f]$ may still increase with r due to the increase of the annulus volume. Note also that $\sigma_t^{(i)}$ are related to fiber orientations ϕ_i of the composite elements in the annuli. For different short-fiber composite material systems, different Weibull shape parameters, m's, are obtainable by various experimental methods. The failure parameter $E[P_f]$ as a function of m is also shown in Fig. 5, where m ranging from 1 to 25 is given for illustrative purposes. The maximum $E[P_f]$ associated with a given m does not always

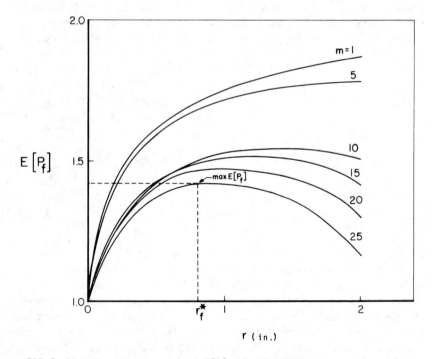

FIG. 5—*Variation of fracture parameter* E[P_f] *with radial distance in composite specimen.*

occur at the center of the panel, where the maximum strain is usually developed. This provides a theoretical basis for explaining the characteristic nature of statistical fracture initiation in the chopped-mat fiber composite—failure caused by the combination of high strain and stress and the most unfavorable flaws.

Variation of r_f^* with the Weibull shape parameter gives a quantitative description of statistical fracture initiation locations in various chopped-mat, fiber-reinforced composites. This is clearly shown in Fig. 6, where the change of r_f^* versus m is presented graphically. As is expected, for a composite with a larger m-value, stress variation is a dominant factor in controlling the local failure. Fracture in this case tends to initiate near the center of the panel. For the specimen with a smaller m, failure tends to initiate well away from the center of the disk due to the statistical size effect. In fact, it is most likely to occur at the boundary of the gage section. For the present glass/epoxy composite material system, the Weibull shape factor m is estimated in the range of 8 to 20. Failure locations for these values of m obtained in the theoretical analysis are found to be consistent with measurements conducted in laboratory experiments [7,8].

Additional Remarks

Because of space limitation, only selected results have been presented to demonstrate the proposed method of approach and to illustrate fundamental characteristics of the statistical fracture initiation behavior in chopped-mat fiber composites. It is clear in the formulation that the local failure of a short-

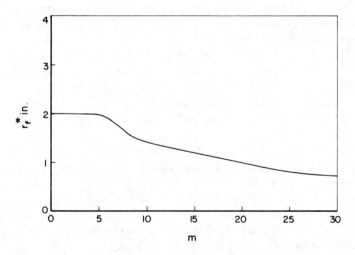

FIG. 6—*Variation of location of fracture initiation* (maximum E[P_f]) *with Weibull shape parameter* m.

fiber composite is directly related to Weibull's scaling, location, and shape parameters of the material. Present solutions address only the effect of the shape parameter on fracture initiation. Furthermore, the two current methods for determining the Weibull parameters of the composite material—that is, the moment estimation and the maximum likelihood estimation—have significant drawbacks in their accuracy [25]. Further investigation is needed to provide more reliable statistical strength parameters for the material so that a more accurate fracture behavior can be obtained.

It is noted that in an actual chopped-mat fiber composite system, an element may have many layers of fiber mats with different orientations through the thickness direction. This complex microstructure requires a fully three-dimensional consideration in modeling the heterogeneous solid. A full-scale Monte Carlo simulation with this more sophisticated model would certainly provide more accurate information and better insight into the problem. This has not been included in the present study.

Summary and Conclusions

Owing to the inherently heterogeneous microstructure and the random fiber orientation, initiation of fracture in chopped-mat, short-fiber composite materials is statistical in nature. It commonly leads to severe, undesirable functional and structural failure of composite structures and components. An analytical approach based on the Weibull statistical failure theory and the concept of mechanics of fracture in brittle fiber composites has been presented. Random orientations of fiber mats and heterogeneity of the composite microstructure are incorporated into the model. The effect of biaxial thermomechanical stress is incorporated into the Weibull form of statistical strength theory in the formulation. A parameter that gives a measure of fracture probability of constituent fiber-mat elements in the composite is introduced. A computer simulation procedure with the aid of a numerical stress analysis has been proposed. For illustration, evaluation of fracture initiation in a statistical sense has been conducted for the disk-type chopped-mat composite subjected to biaxial thermomechanical loading. Selected results have been presented to indicate the fundamental nature of the statistical fracture initiation in the composite.

Based on the study conducted, the following conclusions may be reached:

1. The local microstructure of the randomly oriented chopped-mat, short-fiber composite can be conveniently incorporated into the present analytical model for studying the thermomechanical behavior of this class of material.

2. The multiaxial stress state in the composite introduced by the complex microstructure and loading can be included in the Weibull form of statistical failure theory for studying the current problem.

3. The parameter P_f introduced may be used to determine the fracture probability of constituent elements in the composite. The computer-aided

simulation scheme provides a suitable approach for studying statistical fracture initiation in the chopped-mat fiber composite.

4. Depending upon the value of m and randomness of fiber orientations, failure probability can increase with r even though the stresses in the composite mat elements are found to decrease.

5. Sites of fracture initiation in a statistical sense may be predicted by the proposed scheme in which the combined effect of the worst flaw and the most unfavorable stress in the composite is considered. The predicted fracture initiation location, r_f^*, is found to be sensitive to the Weibull parameters of the material.

Acknowledgments

The authors wish to take this opportunity to acknowledge Dr. J. L. Olinger of the Technical Center, Owens-Corning Fiberglas Corporation, and Professor H. T. Corten of the University of Illinois for their encouragement and valuable discussion during the course of this study.

References

[*1*] Broutman, L. J. and Krock, R. H., *Modern Composite Materials,* Addison-Wesley, Reading, Mass., 1967.

[2] Heimbuch, R. A. and Sanders, B. A., "Mechanical Properties of Chopped Fiber Reinforced Plastics," in *Composite Materials in the Automobile Industry,* S. V. Kulkarni, C. H. Zweben, and R. B. Pipes, Eds., American Society of Mechanical Engineers, New York, 1978, pp. 111-140.

[3] Fesko, D. G., Mallick, P. K., and Newman, S., "Automotive Composites: Manufacturing and Material Interactions," in *Composite Materials in the Automobile Industry,* S. V. Kulkarni, C. H. Zweben, and R. B. Pipes, Eds., American Society of Mechanical Engineers, New York, 1978, pp. 1-10.

[4] Deutsch, G. C., "Automotive Applications of Advanced Composite Materials," *Selective Applications of Materials for Products and Energy,* Vol. 23, Society for Advancement of Materials and Process Engineering, Azusa, Calif., 1978, pp. 34-60.

[5] Irwin, G. R., "Fracture of Pressure Vessels," in *Materials for Missiles and Spacecraft,* McGraw-Hill, New York, 1963, pp. 204-229.

[6] Broek, D., *Elementary Engineering Fracture Mechanics,* Chapter 15, Sijthoff & Noordhoff, The Netherlands, 1978, pp. 366-370.

[7] Socie, D. F., "Characterizing Fatigue Resistance of Fiber Reinforced Plastic in Liquid Natural Gas Pressure Vessels," Technical Report, Owens-Corning Fiberglas Corp., Granville, Ohio, 1978.

[8] Olinger, J. L. and Mott, R. A., "Cryogenic Fiber Reinforced Plastics Development," Intra-Company Technical Report, Technical Center, Owens-Corning Fiberglas Corp., Granville, Ohio, 1978.

[9] Cooper, R. E. and Waters, M. A., *International Journal of Fracture,* Vol. 13, No. 1, 1977, pp. 77-83.

[10] Weibull, W., *Journal of Applied Mechanics,* Vol. 18, No. 3, 1951, pp. 293-297.

[11] Kies, J. A., "The Strength of Glass Fibers and Failure of Filament Wound Pressure Vessels," NRL Research Report 6034, U.S. Naval Research Laboratory, Washington, D.C., 1964.

[12] Scop, P. M. and Argon, A. S., *Journal of Composite Materials,* Vol. 3, 1969, pp. 30-42.

[13] Argon, A. S., "Statistical Aspects of Fracture," Chapter 4, *Composite Materials,* Vol. 5, L. J. Broutman, Ed., Academic Press, New York, 1974, pp. 154-189.

[14] Yang, J. N. and Liu, M. D., *Journal of Composite Materials,* Vol. 12, 1978, pp. 19–38.

[15] Chou, P. C., Wang, A. S. D., Croman, R., Miller, H., and Alper, J., "Statistical Analysis of Strength and Life of Composite Materials," Technical Report AFWAL-TR-80-4049, Air Force Materials Laboratory, Wright-Patterson Air Force Base, Ohio, 1980.

[16] Halpin, J. C., Jerina, K. L., and Johnson, T. A. in *Analysis of Test Methods for High Modulus Fibers and Composites, ASTM STP 521,* American Society for Testing and Materials, 1973, pp. 5–64.

[17] Knight, M. and Hahn, H. T., *Journal of Composite Materials,* Vol. 9, 1975, pp. 77–90.

[18] Coleman, B. D., *Journal of Mechanics and Physics of Solids,* Vol. 7, 1958, pp. 60–70.

[19] Gücer, D. E. and Garland, J., *Journal of Mechanics and Physics of Solids,* Vol. 10, 1962, pp. 365–374.

[20] Zweben, C. and Rosen, B. W., *Journal of Mechanics and Physics of Solids,* Vol. 18, 1970, pp. 189–206.

[21] Oh, K. P., *Journal of Composite Materials,* Vol. 13, 1979, pp. 311–328.

[22] Corten, H.T., TAM 381 class notes, Department of Theoretical and Applied Mechanics, University of Illinois, Urbana, 1979.

[23] "Random Number Generator," UOI-RANBZ-104F-A, IBM 360 Library Routine, CSO Vol. 7, Users Book 1, Digital Computer Laboratory, University of Illinois, Urbana, Ill., 1974.

[24] Richardson, B. C., "Random Number Generation on the IBM 360," Report 329, Department of Computer Science, University of Illinois, Urbana, April 1969.

[25] Talreja, R., "Estimation of Weibull Parameters for Composite Material Strength and Fatigue Life Data," Technical Report 168, The Danish Center for Applied Mathematics and Mechanics, Technical University of Denmark, Lyngby, 1979.

N. S. Sridharan[1]

Elastic and Strength Properties of Continuous/Chopped Glass Fiber Hybrid Sheet Molding Compounds

REFERENCE: Sridharan, N. S., **"Elastic and Strength Properties of Continuous/ Chopped Glass Fiber Hybrid Sheet Molding Compounds,"** *Short Fiber Reinforced Composite Materials, ASTM STP 772,* B. A. Sanders, Ed., American Society for Testing and Materials, 1982, pp. 167–182.

ABSTRACT: Hybrid sheet molding compounds containing both continuous and chopped glass fiber reinforcements have been modeled as laminates of continuous and chopped fiber plies. It is shown that the elastic moduli and the tensile strengths of a series of hybrid compositions with different ratios of continuous to chopped random glass fiber can be calculated from the same constituent plies. Hybrid molding compounds ranging from 100 percent chopped random glass to 90 percent continuous glass fiber at the same total glass content were tested for elastic moduli and ultimate tensile strengths in the axial and transverse directions. Good agreement is obtained with the laminate model calculations. Similar models for other hybrid molding compounds such as XMC materials are developed and the model predictions are compared with data in the literature.

KEY WORDS: composites, elastic properties, glass fiber, laminate analysis, sheet molding compound, tensile strength

The twin pressures of federally mandated fuel economy requirements and the escalating costs of fuel in the 1980s are expected to add impetus to the shift to lightweight structural materials in the transportation industry. One class of materials aimed at this market are sheet molding compounds with continuous glass fibers incorporated in addition to the chopped glass fibers. The continuous fibers contribute to substantial increases in stiffness and strength, though restricted to preferred directions. Ackley [1][2] and Jutte [2] have described processing and preliminary property data on such com-

---

[1]Manager, Composites and Advanced Materials, Materials Engineering and Technology, International Harvester Company, Hinsdale, Ill. 60521.

[2]The italic numbers in brackets refer to the list of references appended to this paper.

pounds. One major stumbling block to the widespread use of these compounds is the lack of a comprehensive database of mechanical properties
needed for the use of advanced design analysis such as finite element
analysis. Denton [3] and Heimbuch and Sanders [4] have characterized high
strength random fiber reinforced systems in detail. Riegner and Sanders [5]
have extended this effort to include selected hybrid sheet molding compounds with both continuous and chopped random glass fibers. This approach is useful and necessary for specific applications. However, the
amount of material data and the time frame over which it is needed make
such a detailed approach impractical and excessively expensive. Figure 1
schematically illustrates the full range of high-strength sheet molding compounds (HSMC) available to the designer. A number of compositions are
possible and optimum weight savings for a given application will depend on
tailoring the material to the application.

A more viable approach is to treat these materials as being composed of
discrete plies of simple reinforcements. The properties of continuous glass
fiber and chopped glass fiber reinforced systems can be combined analytically to develop the properties of the HSMC. This would limit the
characterization effort to a few materials that constitute the discrete plies.
The HSMC with both continuous and chopped random glass fiber reinforcements can be treated as a laminated composite. Micromechanics theory
and laminated plate theory can be used as needed to develop properties for a
material of interest. Such analytical procedures for determining a host of
properties of a laminated composite from those of the constituent plies have

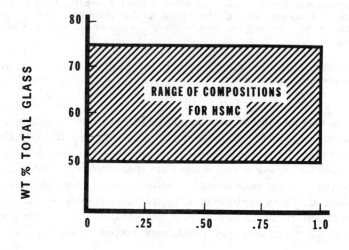

FIG. 1—*Range of commercial HSMCs.*

reached a high degree of sophistication and are routinely used in design and stress analysis [6, 7]. The laminate analogy to represent a complex reinforcement in terms of simpler components has been successfully applied by Halpin and Pagano [8] and Halpin et al [9]. Laminate stiffness calculations have proved very reliable while laminate strength calculations have proved less so. Petit and Waddoups [10] have developed a progressive ply failure criterion to determine the stress-strain response of a laminate from the nonlinear lamina stress-strain curves. Halpin and Kardos [11] have combined the laminate analogy with a linearized progressive ply failure criterion to predict the strength of discontinuous fiber reinforced composite. However, the analysis assumes a "perfectly elastic-plastic" stress-strain curve for the ply in the laminate.

In this study laminate models are constructed to represent commercial HSMC materials and the elastic properties are developed in terms of these models. A linearized progressive ply failure analysis is used to develop upper and lower bounds for laminate strength depending on the stress-strain response of the ply in the laminate. The model predictions are compared with experimental data for both commercial HSMC and special model compounds tested to check the models.

Procedure

Materials

Five HSMC materials were obtained from commercial sources. Table 1 lists the compounds and the sources. The nomenclature used follows that suggested by Riegner and Sanders [12]. The letters "C," "R," and "X" represent continuous, random, and X pattern fibers, respectively. The number following the letter designates the nominal weight percent fibers of that geometry. The wind angle for the X pattern fibers for the HSMC tested was 7.5 deg. Four of these compounds were analyzed for actual continuous and chopped glass fiber content by a method developed by MFG Corporation and subsequently adopted to our needs [13]. The actual glass contents are also shown in Table 1. The matrix in all cases was Dow 790 vinyl ester and the filler where present was calcium carbonate. Two additional materials, an SMC C40R25 and an SMC C15R50, were molded by laminating two commerical HSMCs. Since HSMCs (except for SMC XR) are inherently asymmetric, they were molded in balanced four-ply lay ups. The mold pressure was 7 MPa and the platen temperature was 422 K. The panel dimensions were approximately 450 by 300 mm and the thickness ranged from 2.5 to 5 mm, depending on the compound. Lay ups were chosen to have the chopped fiber at the surface. In addition to unidirectional lay ups, $\{\pm 18\}_s$ and $\{\pm 45\}_s$ panels were also molded in the case of SMC CR materials. Axial and transverse samples were prepared from each panel after discarding a 13-mm-

TABLE 1—*HSMCs tested and actual glass contents.*

Material	Source[a]	Weight Percent Continuous Fiber	Weight Percent Chopped Glass Fiber
SMC C60R5	OCF	62	6
SMC C45R20	OCF	49	17
SMC C30R35	OCF	31	36
SMC R65	OCF	0	67
SMC 7.5X50R25	PPG	50	26

[a]OCF = Owens-Corning Fiberglas Corporation, Granville, Ohio; PPG = PPG Industries, Pittsburgh, Pa.

wide strip along the periphery. The samples were tabbed in sections using a glass fiber reinforced circuit board material (Cimbord 2FR from Cincinnati Milacron). Straight-sided specimens 230 by 25 mm were cut out of the tabbed sheet using a diamond saw. The tabs were approximately 38 mm long and had a 30-deg taper angle. The adhesive was a commercial cyanoacrylate (Loctite 414).

Experiments

The specimens were tested for tensile modulus and ultimate tensile strength in accordance with the ASTM Test for Tensile Properties of Fiber-Resin Composites (D 3039) in an Instron universal testing machine. The crosshead rate was 3 mm/s. The axial and transverse strains were recorded with extensometers with an effective gage length of 25 mm. The number of specimens ranged from 5 for the model materials to 15 for the commercial HSMCs. The elastic moduli and the ultimate tensile strengths were determined in the axial and transverse directions for each material and lay up. The mean values with the associated standard deviations are shown in Table 2.

Laminate Analogy

The HSMC materials are composed of continuous and chopped glass fiber in the same matrix. A critical assumption is that local fiber fraction of the HSMC is uniform and constant for both the continuous and chopped glass fibers. This assumption allows a unique definition of a laminate analog for each HSMC to reflect the geometry and material balance. The relative thickness of the plies containing continuous or chopped fiber is proportional to the weight fraction of that fiber in the HSMC to satisfy the material balance requirement. The geometry of the SMC CR material is directly reflected in the laminate analog as the continuous and chopped fiber are distributed in preferred planes in the material. This requirement is more complex in the case of HSMC with an X pattern to the continuous fiber. In

TABLE 2—*Tensile moduli and strengths of HSMCs.*

Material	Lay Up	Number of Specimens	Tensile Modulus, GPa		Tensile Strength, MPa	
			Mean	Standard Deviation	Mean	Standard Deviation
SMC C60R5	{0}	15	37.9	2.0	727.4	34.5
	{90}	15	9.5	1.3	28.3	2.8
	{±18}s	5	36.3	2.8	288.8	20.7
	{±45}s	5	16.7	0.3	120.9	2.8
	{±72}s	5	12.1	1.4	28.2	2.8
SMC C45R20	{0}	15	33.6	2.8	682.6	34.5
	{90}	15	13.0	1.4	55.2	3.4
	{±18}s	5	23.7	1.4	243.3	6.7
	{±45}s	5	16.3	1.6	182.0	17.9
	{±72}s	5	10.9	0.9	71.8	6.1
SMC C30R35	{0}	5	27.1	0.5	459.2	27.4
	{90}	5	12.6	0.6	102.4	0.4
	{±18}s	5	22.3	0.3	245.5	9.9
	{±45}s	5	13.8	0.6	132.8	15.2
	{±72}s	5	12.4	0.6	106.2	0.4
SMC C40R25	{0}	5	28.3	1.1	549.0	6.8
	{90}	5	11.8	0.8	89.9	4.2
SMC C15R50	{0}	5	20.7	0.3	325.7	21.4
	{90}	5	15.7	0.3	170.2	14.3
SMC R65	{0}	15	18.0	0.6	245.0	6.4
	{90}	15	15.7	0.9	205.7	11.0
SMC 7.5X50R25	{0}	15	38.3	1.6	578.7	21.0
	{90}	15	16.5	1.0	72.1	5.3

these HSMCs the chopped fiber is fully interspersed with the continuous fiber. To account for the geometry of the material as well as the inherent symmetry, the material is modeled as a six-ply laminate with $\{\pm 7.5\}_s$ laminate continuous fiber sandwiched between plies containing the chopped fiber. The fiber weight fraction in all cases is equal to the total fiber weight fraction of the HSMC. For example, the laminate analog for an SMC C60R5 consists of C65 and R65 plies, with the C65 plies being twelve times as thick as the R65 plies. The laminate analogs for all the materials tested are shown in Table 3.

Calculation of Elastic Properties

For determining the elastic properties of the HSMC from the laminate analog, the in-plane elastic properties of the constituent plies (C65, R65, C75, and R75) are needed. Since experimental data were available only for the R65 material, the elastic properties were determined by a micromechanics analysis developed by McCullough et al [14]. This was available in the form of a computer program developed in the Center for Composite Materials at the University of Delaware. In addition to the elastic properties of the fiber, filler, and resin (Table 4), fiber and filler fraction and fiber orientation are needed to calculate the laminate properties. The values used to define the constituent plies and the resultant in-plane elastic properties are shown in Table 5. The only parameters not fixed in terms of the composition for the HSMC are the orientation and fiber aspect ratio for the chopped fiber component of the SMC X50R25 material. These values were estimated to be 0.6 and 50, respectively, from the best of the measured axial and transverse moduli of the material. The deviation of the actual continuous and chopped glass fiber contents from the nominal values, shown in Table 1, were ignored in the analysis.

The data in Tables 3 and 5 are totally sufficient for determining the elastic properties of the HSMC from the laminate analog using classical laminated plate theory. By suitable transformation of the stiffness matrices, an effective elastic modulus for the laminate analogue can be calculated. This procedure has been described in detail elsewhere [11,15]. Although specific cases such as the moduli of SMC CR materials in fiber direction can be handled more simply using rule-of-mixtures, laminated plate theory is more general and can handle more complex systems such as SMC XR materials and angle ply lay ups of SMC CR material.

Calculation of Tensile Strength

The strengths were determined by a progressive ply failure analysis. A failure criterion and the post-failure behavior of the ply in the laminate need to be defined. A maximum strain ply failure criterion [10] was chosen. It

TABLE 3—*Laminate analogs for HSMCs in a direction θ.*

HSMC	Model	Continuous Fiber Ply		Chopped Fiber Ply	
		Composition[a]	Thickness[b]	Composition[a]	Thickness[b]
SMC C60R5	$\{\theta^c/\theta^R\}$	65	0.093	65	0.007
SMC C45R20	$\{\theta^c/\theta^R\}$	65	0.069	65	0.031
SMC C40R25	$\{\theta^c/\theta^R\}$	65	0.062	65	0.038
SMC C30R35	$\{\theta^c/\theta^R\}$	65	0.046	65	0.054
SMC C15R60	$\{\theta^c/\theta^R\}$	65	0.023	65	0.077
SMC X50R25	$\{(\theta \pm 7.5)^c/\theta^R\}_s$	75	0.017	75	0.032

[a]Percent by weight of glass fiber.
[b]Based on a nominal laminate thickness of 0.100.

TABLE 4—*Elastic properties of HSMC components.*

Material	Young's Modulus, GPa	Shear Modulus, GPa	Poisson's Ratio
Vinyl ester	3.52	1.31	0.301
E-glass fiber	72.4	27.2	0.333
Calcium carbonate	47.78	18.06	0.023

must be noted that for the SMC CR material in the fiber and transverse directions, maximum stress and maximum strain will yield identical results in a linear analysis with the proper choice of critical strains or stresses. Another critical question is the load redistribution in the laminate after failure of each ply until total laminate failure. In their analysis of the strength of discontinuous fiber composites, Halpin and Kardos [11] assumed that the compliances of the failed ply are infinite while the load remains unchanged until laminate failure. This implies a ply stress-strain curve which is "perfectly elastic–plastic", as shown schematically in Fig. 2a. This is not a realistic assumption for a quasi-brittle system. Figure 2b illustrates the other extreme, where the failed ply unloads totally and does not carry any load. This probably reflects the situation in hybrids with substantial differences in ply stiffness and failure strain together with poor interlaminar properties. Hybrids of S-glass and graphite discussed by Kalnin [16] will probably fall in this category. For the HSMC systems discussed in this paper, Fig. 2c probably illustrates the real situation, where the failed ply continues to support load because of shear transfer and this fraction diminishes on further loading of the laminate. The first two situations represent the upper and lower bounds. In the absence of detailed information on damage propagation, these bounds were calculated for SMC CR materials in this study.

The upper and lower bounds of the strength of SMC CR material in tension are calculated as shown below.

In the axial direction:

$$\sigma_A \text{ (upper bound)} = E_{11}\epsilon^R + E^*_{11}(\epsilon_1{}^c - \epsilon^R) \tag{1}$$

$$\sigma_A \text{ (lower bound)} = \text{lesser of } E_{11}\,\epsilon^R \quad \text{or} \tag{2}$$
$$E^*_{11}\,\epsilon_1{}^c \tag{3}$$

Equation 2 describes the situation where the first ply failure constitutes laminate failure.

In the transverse direction:

$$\sigma_T \text{ (upper bound)} = E_{22}\,\epsilon_2{}^c + E^*_{22}(\epsilon^R - \epsilon_2{}^c)$$

TABLE 5—*In-plane elastic properties of constituent plies.*

Ply	Composition[a]				Fiber Geometry		Elastic Properties[c]			
	Fiber	Resin	Filler	Aspect Ratio	Orientation		E_{11}, GPa	E_{22}, GPa	ν_{12}	G_{12}, GPa
C65	67[b]	30	3	10^6	1.0		39.8	10.34	0.300	4.48
R65	67[b]	30	3	50	0.0		18.5	18.50	0.289	7.20
R75	75	25	0	50	0.6		33.45	17.72	0.393	9.70
C75	75	25	0	10^6	1.0		42.06	15.17	0.326	6.90

[a]Weight percent.
[b]From the average of measured glass contents.
[c]E = Young's modulus, ν = Poisson's ratio, and G = shear modulus.

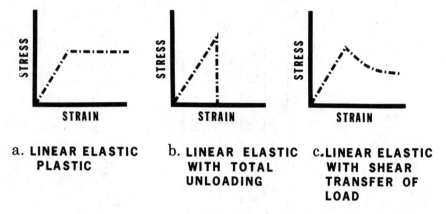

a. LINEAR ELASTIC
   PLASTIC

b. LINEAR ELASTIC
   WITH TOTAL
   UNLOADING

c. LINEAR ELASTIC
   WITH SHEAR
   TRANSFER OF
   LOAD

FIG. 2—*Schematic illustration of ply stress-strain curves in the laminate.*

where

σ_A, σ_T = tensile strengths in axial and transverse directions, respectively,

E_{11}, E_{22} = effective tensile moduli of the laminate in the axial and transverse directions, respectively,

E^*_{11} = effective axial modulus after random ply failure,

E^*_{22} = effective transverse modulus after continuous fiber ply failure,

ϵ_1^c and ϵ_2^c = allowable strains for the continuous fiber ply in the axial and transverse directions, respectively, and

ϵ^R = allowable strain for the chopped fiber reinforced ply.

It should be noted again that, for the SMC CR materials, in the material directions the upper bound value will reduce to that obtained by a rule-of-mixtures analysis. The three allowable strain values were calculated from the axial and transverse strengths of R65 and C60R5 materials. The allowable strains for the continuous fiber ply were 0.021 and 0.0013 in the axial and transverse directions, respectively. The strain allowable for the R65 material was 0.013 for both directions.

Results

The measured values of elastic moduli for SMC CR materials in the axial and transverse directions are compared with the laminate model calculation in Fig. 3. The formulations all contained 65 percent by weight of glass fiber with the continuous fiber fraction ranging from 0 to 0.9 percent of the total glass content. The measured and calculated values of the effective axial

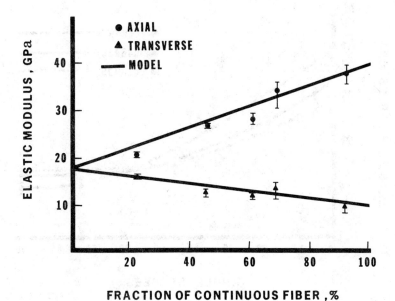

FIG. 3—*Axial and transverse effective tensile moduli as a function of fraction of continuous glass fiber for SMC CR materials with 65 percent by weight glass fiber.*

modulus for angle ply hybrids of SMC C60R5, SMC C45R20, and SMC C30R35 are compared in Fig. 4. In a similar comparison (Fig. 5) for a material with a more complex reinforcement geometry, the effective modulus of off-axis specimens of SMC 7.5X50R25, from the work of Riegner and Sanders [5], is compared with laminate theory calculations. The model calculations are based on the six-ply laminate analog discussed in an earlier section. In general the agreement between the measured and calculated values is very good.

As expected, the fit between measured and calculated data is not as good in the case of tensile strength data. Figure 6 shows a comparison of the tensile strength of SMC CR materials containing 65 precent by weight total glass fiber with the upper and lower bound calculations from the ply failure analysis (Eqs 1 and 2). The results appear to lie closer to the upper bound value. This is not necessarily so, as minimal changes in the values of strain allowables used could change the situation considerably. Figure 7 shows reasonable agreement between the measured and calculated values of transverse strengths of SMC CR materials with 65 percent by weight glass fiber. Only the upper bound is shown, as the difference between upper and lower bounds was not significant with respect to the errors of the experiment. Given the inherent variability of the HSMC materials in strength, the progressive ply failure analysis provides useful estimates of tensile strength

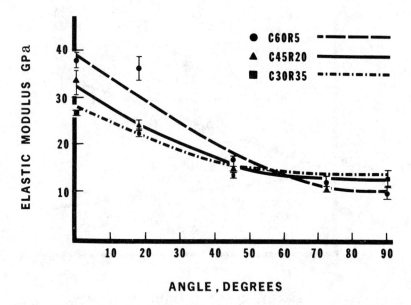

FIG. 4—*Variation of effective tensile modulus for three SMC CR compounds with 65 percent by weight glass fiber.*

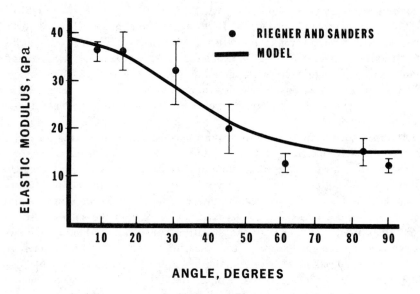

FIG. 5—*Variation of effective tensile modulus of SMC 7.5X50R25 as a function of angle for off-axis specimens.*

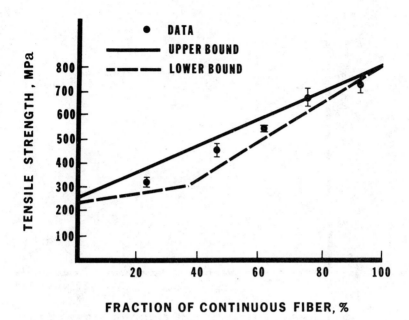

FIG. 6—*Tensile ultimate strength as function of fraction of continuous glass fiber for SMC CR materials with 65 weight percent glass fiber.*

in the material directions. Although the calculation can be extended to angle ply materials or SMC XR compounds, the strain allowables were not available for a meaningful comparison between experimental data and the models.

Discussion

Elastic Moduli of HSMCs

The analytical approach described is quite straightforward and can be applied to the full range of HSMC compositions indicated in Fig. 1. Although laminated plate analysis may not be necessary for several specific cases, it is more general and provides a consistent basis for analytical prediction of moduli. The secondary benefit is that the analysis can be used to determine the full range of elastic properties. One significant fallout from the analysis is that the HSMC is very efficient in utilizing the stiffness of the fiber in the composite. The major assumption in the model is that the fiber and resin are uniformly distributed. Figure 8 is a micrograph of the cross section of an SMC C60R5 material. The preferred planes of distribution of the continuous

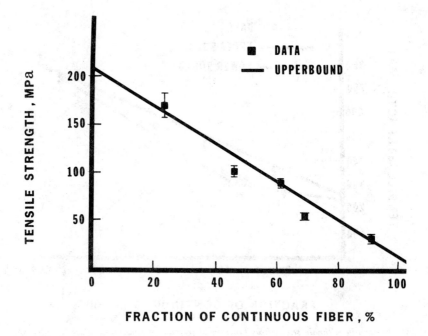

FIG. 7—*Transverse tensile strength as a function of fraction of continuous glass fiber for SMC CR materials with 65 percent by weight glass fiber.*

and chopped glass fiber are clearly seen. The ratio of the thicknesses of the two layers is also of the order of the model parameters. However, in a more detailed study in progress [*17*], it was found that substantial variation of the layer thicknesses, meandering of the random and continuous fiber layers, and nonuniform distribution of glass on a local level are more the rule than the exception. It is to be concluded that the stiffness of the HSMC is not sensitive to these local variations.

Strength of HSMC

The analysis as developed is restrictive and it is possible to develop a more refined analysis that incorporates nonlinear stress-strain behavior or interactive failure criteria [*18*]. However, the measured properties are more sensitive to the inherent variability in the material and the processing than the predicted values are to the assumptions. The analysis provides a basis for calculating reasonable estimates of strength, albeit only in the fiber and transverse directions, for a series of HSMCs at the same total glass fiber fraction as the ratio of continuous to chopped glass fiber is varied. To extend the analysis to the full range of HSMCs shown in Fig. 1, critical stresses or strains must be determined for a number of additional model materials.

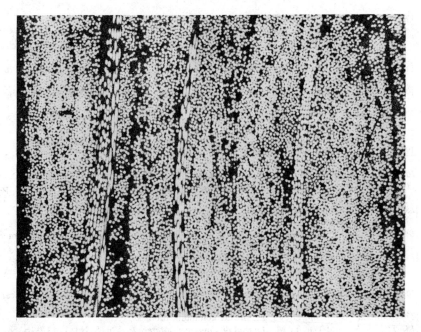

FIG. 8—*Cross section of SMC C60R5 material indicating continuous and chopped random glass fiber layers.*

Conclusions

Hybrid sheet molding compounds containing both continuous glass fiber and chopped glass fiber have been modeled as laminates of plies with the individual reinforcements. Analytical predictions of the elastic moduli based on the laminate model and experimental determination of the values for various HSMC compositions show good agreement. Strength calculations based on the same model provide usable estimates of tensile strength for SMC CR materials. The HSMCs are very efficient in utilizing the stiffness of the reinforcing fiber in the material in spite of substantial variability in the microstructure.

Acknowledgments

I wish to thank J. G. Claussen and S. Dichter for help in acquiring the compounds and data analysis, M. L. Hinduja for processing assistance, and D. B. Edwards for much of the testing.

References

[1] Ackley, R. H. and Carey, E. P. in *Proceedings*, 34th SPI Conference, 1979.
[2] Jutte, R. B., Paper No. 780355, Society of Automotive Engineers, 1978.

[3] Denton, D. L. in *Proceedings*, 34th SPI Conference, 1979.
[4] Heimbuch, R. A. and Sanders, B. A., "Mechanical Properties of Automotive Chopped Fiber Reinforced Plastics," Report MID-78-032, General Motors Corp., Warren, Mich., 1978.
[5] Riegner, D. A. and Sanders, B. A., presented at the National Technical Conference of the Society of Plastics Engineers, Detroit, 1979.
[6] Ashton, J. E., Halpin, J. C., and Petit, P. H., *Primer on Composite Materials: Analysis*, Technomic Publishing Co., Stamford, Conn., 1969.
[7] Ashton, J. E. and Whitney, J. M., *Theory of Laminated Plates*, Technomic Publishing Co., Stamford, Conn., 1970.
[8] Halpin, J. C. and Pagano, N. J., *Journal of Composite Materials*, Vol. 3, 1969, p. 720.
[9] Halpin, J. C., Jerina, K. L., and Whitney, J. M., *Journal of Composite Materials*, Vol. 5, 1971, p. 36.
[10] Petit, P. H. and Waddoups, M. E., *Journal of Composite* Materials, Vol. 3, 1969, p. 2.
[11] Halpin, J. C. and Kardos, J. L., *Polymer Engineering and Science*, Vol. 18, No. 6, 1978, p. 496.
[12] Riegner, D. A. and Sanders, B. A., in *Proceedings*, 35th SPI Conference, 1980.
[13] "SMC Technical Review," Owens-Corning Fiberglas Corporation, Granville, Ohio.
[14] McCullough, R. L., Jarzebski, G., McGee, S., and Mroz, P., Internal Report of the Center for Composite Materials, University of Delaware, Newark, 1979.
[15] Wetherhold, R. C. and Pipes, R. B., "CMAP-1: Composite Material Analysis of Plates," Internal Report of the Center for Composite Materials, University of Delaware, Newark, 1979.
[16] Kalnin, I. L. in *Composite Materials: Testing and Design*, ASTM STP 497, American Society for Testing and Materials, 1971, pp. 551-563.
[17] Sridharan, N. S. and Drews, T., private communication.
[18] Tsai, S. W. and Wu, E. M., *Journal of Composite Materials*, Vol. 5, No. 1, 1970, p. 58.

Jon Collister[1] and Michael Gruskiewicz[1]

Dynamic Mechanical Characterization of Fiber Filled Unsaturated Polyester Composites

REFERENCE: Collister, Jon and Gruskiewicz, Michael, **"Dynamic Mechanical Characterization of Fiber Filled Unsaturated Polyester Composites,"** *Short Fiber Reinforced Composite Materials, ASTM STP 772,* B. A. Sanders, Ed., American Society for Testing and Materials, 1982, pp. 183-207.

ABSTRACT: Dynamic mechanical testing was used to qualitatively study the effects of various constituents of sheet molding compound (SMC) and bulk molding compound (BMC) materials, including unsaturated polyester resin, a chemical thickener (magnesium oxide), a filler (calcium carbonate), and glass fibers. Tests were designed to evaluate the influence of each constituent as it was added to the overall composition. The parameters tested were elastic shear modulus, viscous shear modulus, and complex viscosity as functions of frequency for non-cross-linked materials, and elastic shear modulus, viscous shear modulus, and damping as functions of temperature for cross-linked materials. The chemical thickener was observed to change the molecular weight of the polyester resin, as indicated by the appearance of a rubbery plateau in the viscoelastic spectrum, while cross-linking, filler, and glass fibers were observed to contribute heavily to the elastic shear modulus and temperature sensitivity.

KEY WORDS: dynamic mechanical properties, unsaturated polyester resin, glass fiber reinforcement, mineral filler, chemical thickening, shear moduli, complex viscosity, sheet molding compounds, bulk molding compounds

Sheet molding compounds (SMC) and bulk molding compounds (BMC) are complex composite materials comprised of unsaturated polyester resins, unsaturated monomers, thermoplastic polymers, mineral fillers, glass fibers, chemical thickeners, mold release agents, colorants, polymerization initiators, and polymerization inhibitors. When formulated and molded properly, these constituents yield composite materials that have a relatively low cost while exhibiting very respectable mechanical properties, electrical properties, corrosion resistance, and stiffness-to-weight ratio. These characteristics, along

[1]R&D Manager and Research Associate/Rheology, respectively, Premix, Inc., North Kingsville, Ohio 44068.

with the ease in molding complex geometries, have given SMC and BMC wide industrial acceptance and usage in a number of market applications.

The unsaturated polyester polymers used in SMC and BMC go through several significant changes as they are processed into the final molding. Depending on the molecular weight and chemical composition, the polymers may be glassy or fluid materials at room temperature. When dissolved in styrene (the most commonly used unsaturated monomer) they become Newtonian fluids that can then be easily handled for shipping and storage purposes. When compounded with the fillers, fibers, and other ingredients, a complex suspension is achieved that is decidedly non-Newtonian in behavior. After the addition of chemical thickening agents, the viscosity may rise by orders of magnitude, depending on the concentration of thickening agent. This viscosity rise is theorized to be a result of an increase in the molecular weight of the polyester polymer due to formation of double polyester salts of the metallic ion [1].[2] The viscosity rise is necessary to give the suspension sufficient elasticity to prevent glass fiber separation during mold flow. This material is then molded and cross-linked to produce a glassy material several orders of magnitude higher in elastic modulus.

From this discussion, it becomes obvious that to completely understand the processing and utility of these materials, one must study the material in all stages of development. Dynamic mechanical testing is well suited for this purpose.

The advantages of dynamic mechanical testing for these materials are numerous. Normally it is extremely difficult to test SMC and BMC in the fully formulated uncross-linked state, due to the effects of glass fibers. Rotational viscometers become useless when fibers are present. Additionally, even when fibers are not present the magnitude of the changes caused by fillers and chemical thickeners necessitate the use of several spindle sizes and angular velocities. Measurements made with differing shear rates are not useful for comparisons in non-Newtonian materials.

For cross-linked samples, much more is known due to the facility in testing solid materials. Static mechanical, electrical, corrosion, fatigue, and creep properties are widely known for several SMC and BMC formulations. Dynamic mechanical testing offers insight into structural transitions within the cross-linked polymer that are difficult to investigate with other methods. Also, since the method is nondestructive (or relatively so) one sample can be used to generate a large quantity of information such as moduli versus temperature experiments.

It is therefore the object of this paper to present the data gathered from fluid and solid dynamic mechanical testing of an SMC formulation at various stages. It is intended that a qualitative view of the role of various constituents be demonstrated.

---

[2]The italic numbers in brackets refer to the list of references appended to this paper.

Experimental Procedure

Materials

The unsaturated polyester resin used in this study was a commercially available general purpose resin comprised of isophthalic acid, maleic anhydride, and propylene glycol. The chemical thickening agent was an 85 percent pure grade of magnesium oxide. The filler material was calcium carbonate (3 micron average diameter). The glass fibers were 6.35-mm prechopped fibers with 2 percent organic binder of which 15 percent is styrene soluble. For the cross-linked specimens, organic peroxides were used to initiate the polymerization. Zinc stearate was added as a mold release. The formulations tested are shown in Table 1.

Instrumentation

The instrument used in these experiments was a Rheometrics Thermal Mechanical Spectrometer,[3] which is a forced torsion oscillating rheometer. A sinusoidal deformation of controlled amplitude and frequency is applied to the sample. The input stress wave is altered as it passes through the sample depending on the viscoelastic state of the material. A transducer then detects the output stress transmitted through the sample, and a microprocessor determines the phase angle difference between the stress and strain. The relationship is shown in Fig. 1 [2]. This relationship can also be represented with rotating vector diagrams (Fig. 2) [2].

The output vector $G*$ is resolved into components: $G' = G*$ cosine δ, an elastic shear modulus in phase with the applied stress; and $G'' = G*$ sine δ, the viscous shear modulus 90 deg out of phase with the stress. It is seen that

$$G* = [(G')^2 + (G'')^2]^{1/2}$$

A complex viscosity ($\eta*$) is also defined such that [2]

$$\eta* = \frac{[(G')^2 + (G'')^{1/2}}{\omega}$$

where ω is defined as angular frequency and is expressed in radians per second (rad/s).

Damping is defined as the tangent of the phase angle δ as follows:

$$\tan \delta = G''/G'$$

[3]Rheometrics, Inc., Union, N.J.

TABLE 1—*Formulations tested.*[a]

	1	2	3	4	5	6	7
Polyester	100	70	70	70	70	70	70
Styrene	—	30	30	30	30	30	30
MgO	—	2.3	—	1.5	1.5	1.5	1.5
Peroxide	—	—	1.0[b]	1.0[c]	1.0[c]	1.0[c]	1.0[c]
CaCO$_3$	—	—	—	150	200	150	150
Zinc stearate	—	—	—	3.8	3.8	3.8	3.8
Glass fibers	—	—	—	—	—	28(10%)	64(20%)

[a]All values are given in parts per hundred resin (phr).
[b]Methyl ethyl ketone peroxide (85 percent pure).
[c]Tertiary butyl perbenzoate.

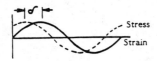

FIG. 1—*Phase angle difference between stress and strain.*

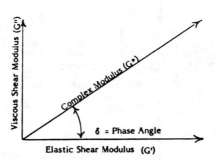

FIG. 2—*Rotating vector diagrams.*

The test geometry for the non-cross-linked fluid materials is shown in Fig. 3. For the materials of low viscosity, parallel plates with a 50 mm diameter were used. For the higher viscosity materials, 25 mm parallel plates were used. The height h was 2 mm in all cases. The maximum angular displacement θ was adjusted to yield 2.5 percent strain.

The test geometry for the solid testing was a rectangular beam approximately 46 by 3.2 by 12.7 mm. This beam was gripped at either end and an oscillatory torque was applied (Fig. 4). The value of θ was chosen to apply an 0.1 percent maximum strain to the sample.

The testing was performed for each sample as follows:

1. *Polyester alkyd*—G', G'', and tan δ versus T (temperature in degrees centigrade) for the glassy region, and G', G'', and η^* at 100°C in the melt.

2. *Thickened polyester resin*—G', G'', and η^* versus ω at 25°C at 5, 26, and 892 h after the addition of magnesium oxide.

3. *Cross-linked polyester resin*—G', G'', and tan δ versus T at 1 Hz.

4 and 5. *Filled polyester compound*—(4) G', G'', and η^* versus ω at 25°C for the non-cross-linked samples. (5) G', G'', and tan δ versus T for the cross-linked samples.

6 and 7. *Filled reinforced polyester compound*—(6) G', G'', and η^* versus ω for the non-cross-linked samples. (7) G', G'', and tan δ versus T for the cross-linked samples.

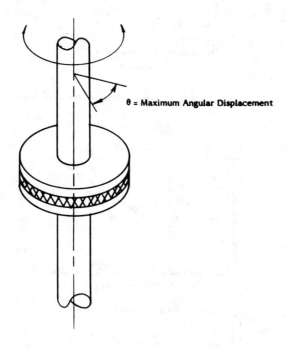

θ = Maximum Angular Displacement

FIG. 3—*Test geometry for non–cross-linked fluid materials.*

Results and Discussion

Polyester Alkyd

The solids testing of the alkyd resin at low temperature is presented in Fig. 5. As can be observed, the polyester alkyd is a glassy solid with G' in the 10^9 dyne/cm^2 decade. The polyester shows a prominent secondary transition, as evidenced by the maxima in G'' and tan δ, which is centered at approximately $-100°C$. At room temperature the polyester alkyd begins to soften; it enters the melt region at approximately 60°C. The polyester exhibits no rubbery plateau; this is typical of low molecular weight materials without sufficient molecular length to exhibit entanglement coupling.

Figure 6 shows the polyester alkyd in the melt at 100°C. The polyester resin at this temperature exhibits Newtonian fluid–like behavior, as evidenced by the insensitivity of η^* to ω. One also notices that G' has a higher slope than G'', as expected from terminal zone viscoelastic theory as developed by Rouse [3].

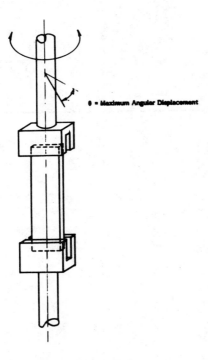

θ = Maximum Angular Displacement

FIG. 4—*Test geometry for solid testing.*

Thickened Polyester Resin

The change in viscosity of the polyester resin upon thickening with MgO is quite dramatic. The change in G', G'', and η^* for one polyester resin sample with magnesium oxide is demonstrated in Figs. 7 to 9 for 5, 26, and 892 h after the addition of magnesium oxide, respectively. Initially an attempt was made to measure the properties of the unthickened polyester resin, but most of the properties lie below the 10^3 dyne/cm^2 decade. This is below the detection limit of the transducer used in this experiment; therefore, we started the series in time at 5 h. One can see that the material exhibits terminal region behavior at 5 h (Fig. 7), with the viscosity being Newtonian over a portion of the frequency range and only slightly non-Newtonian at the higher frequencies. At 26 h (Fig. 8), one can see that the properties have risen an order of magnitude and G'' and G' intersect at the higher frequency. The parameter η^* definitely exhibits non-Newtonian behavior over most of the frequency range.

At 892 h (Fig. 9) one can see a definite development of the plateau in G'

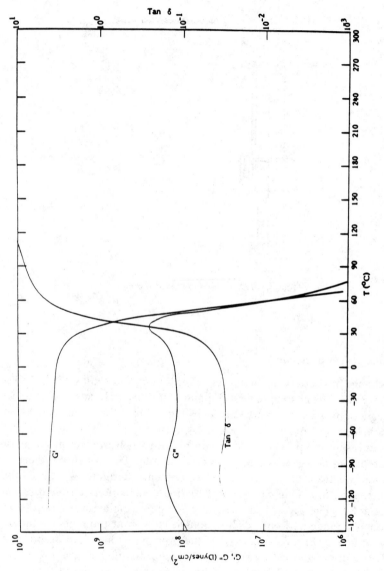

FIG. 5—G', G", and tan δ versus T for non-cross-linked polyester alkyd.

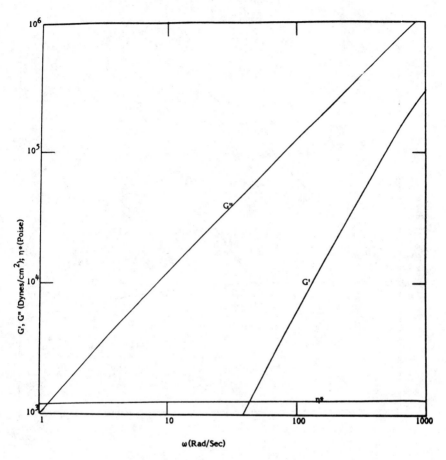

FIG. 6—*G', G", and η* versus ω for polyester alkyd melt at 100°C.*

and the minimum in G'', while η^* is exhibiting non-Newtonian shear thin-ning behavior over the entire frequency range tested.

This trend with the viscoelastic properties demonstrating a plateau zone in G' and a minimum in G'' is very characteristic of polymers that have suffi-cient molecular length to exhibit entanglement coupling. The generalized response of elastic shear modulus is shown in Fig. 10, which demonstrates that low molecular weight materials do not exhibit a rubbery plateau due to insufficient molecular length. Polymers that are above the critical molecular weight show a definite rubbery plateau and terminal region at much lower frequencies [4].

Polyester resins in general [1,5] and this polyester resin specifically [6] have been demonstrated to increase in molecular weight when mixed with

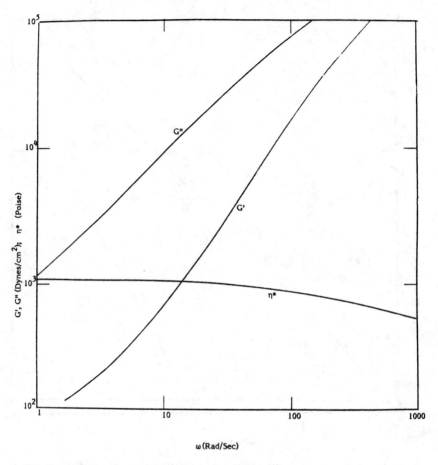

FIG. 7—*G′, G″, and η* versus ω for magnesium oxide thickened polyester resin at 5 h.*

magnesium oxide. It is theorized that doubly acid terminated polymer species can form extended molecular weight polyester salts that then exhibit sufficient molecular length for entanglement coupling. Test results using gel permeation chromatography (GPC) as the measurement of molecular weight have demonstrated that only a fraction of the polyester resin increases in molecular weight. This fraction of high molecular weight polyester salt may be responsible for the dramatic changes that are noticed in the plateauing of $G′$ and the minimum observed in $G″$. Recent data from Soong, Shen, and Hong [7] on blends of polystyrene of various molecular weights demonstrate that even minor blends of high molecular weight material with a low

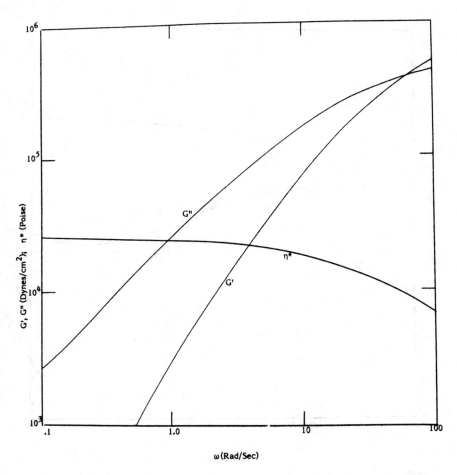

FIG. 8—G', G'', and η^* versus ω for magnesium oxide thickened polyester resin at 26 h.

molecular weight polystyrene dramatically change the behavior of G' and G''. These data for G' are demonstrated in Fig. 11, which shows the plateau-ing of G' versus ω. The corresponding data exhibiting a minimum in G'' ver-sus ω are demonstrated in Fig. 12.

Earlier work has shown that these properties are also a function of the con-centration of thickening agent used in the formulation [8]. This characteris-tic is remarkable in that an infinite variety of viscoelastic states are available to the compounders of SMC to suit individual molding applications. This is one of the unique features of SMC that makes the rheological characteriza-tion of these materials so complex.

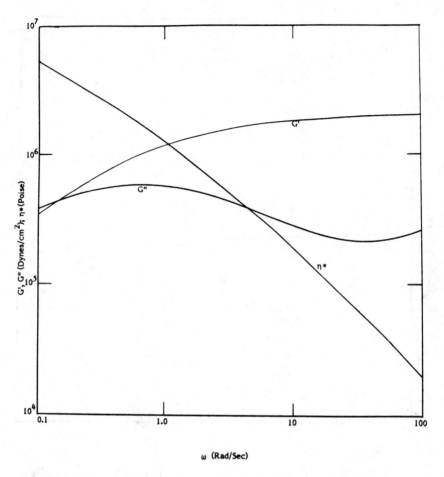

FIG. 9—*G', G", and η* versus ω for magnesium oxide thickened polyester resin at 892 h.*

Non-Cross-Linked Filled Polyester Compound

Figures 13 and 14 show G', G'', and η^* versus ω at room temperature for the samples containing 150 and 200 parts per hundred resin (phr) $CaCO_3$ respectively. In these plots we see the same tendency of the thickened polyester to show an intersection of G' and G'' with G' just starting to exhibit a plateau region. The complex viscosity is non-Newtonian throughout the frequency range tested; however, there is a good development of the terminal zone properties of G'' and G'.

All three properties measured were noticeably higher than for the unfilled polyester resin, but no distortion of the general characteristics was noticed. A

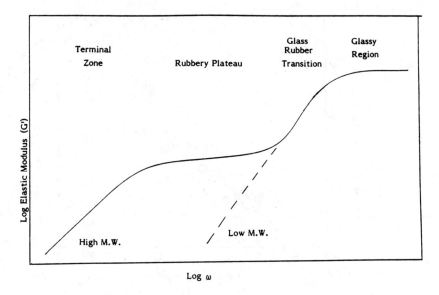

FIG. 10—*Generalized response as elastic modulus versus frequency for high and low molecular weight polymers.*

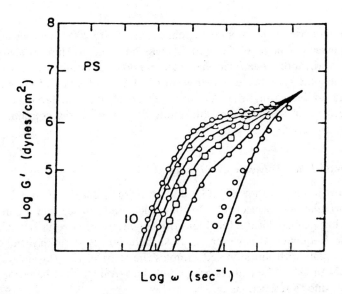

FIG. 11—*G' versus frequency for polystyrene blends. The two outer curves are for pure components; the others represent blends of the pure components comprising 20:80, 40:60, and 80:20 percent by weight. After Ref 7; reprinted by permission of the* Journal of Rheology.

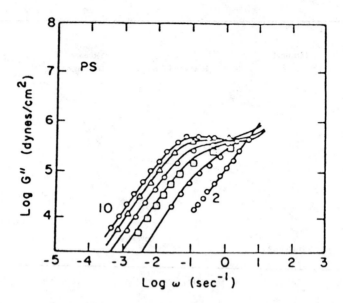

FIG. 12—*G" versus frequency for polystyrene blends. The two outer curves are for pure components; the others represent blends of the pure components comprising 20:80, 40:60, and 80:20 percent by weight. After Ref 7; reprinted by permission of the* Journal of Rheology.

definite increase in the properties going from 150 to 200 parts was observed. The intersection point of G'' and G' was observed to shift to a lower frequency with additional filler material. Pervious unpublished investigations by this laboratory indicate that this shift in the intersection of G' and G'' is not due to increasing the volume fraction of filler material, but is an enhancement of the degree of thickening due to impurities contained within the filler material.

Reinforced Filled Polyester Compound

Figures 15 and 16 show G', G'', and $\eta*$ versus ω for the 10 and 20 percent glass formulations, respectively. In these plots one can see that a generalized shape characteristic of the filled and thickened polyester resin is retained. At the low frequencies, however, a yield behavior is definitely noticed. In previous plots with unreinforced samples the slope of G' was always greater than that of G''. When fibers are added the terminal zone behavior is most effected, due to restrictions on viscous flow.

One can see that the addition of glass fibers does raise the overall properties over the filled and unfilled polyester resin. It does not appear to dramatically alter the intersectin of G' and G''.

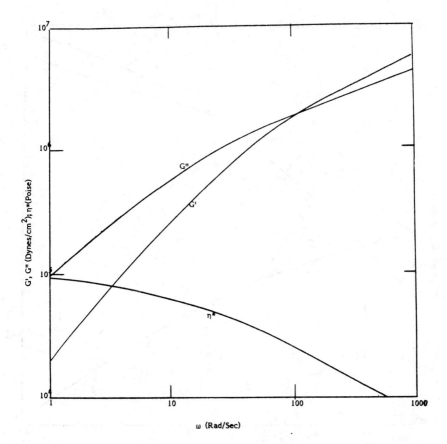

FIG. 13—G', G'', and η^* versus ω for calcium carbonate filled (150 phr) thickened polyester resin.

Cross-Linked Polyester Resin

In Fig. 17, G', G'', and tan δ are plotted versus T for the cross-linked polyester resin. In this plot a secondary transition around $-100°C$ is noticed that is similar to the one noticed in Fig. 5 for the alkyd polyester. The polyester resin softens considerably above $60°C$ and shows a maximum in the damping curve (tan δ) at $170°C$. The polyester resin is entering a plateau region in G' at the upper temperatures tested.

Cross-Linked Filled Polyester Compound

Figures 18 and 19 show G', G'', and tan δ versus temperature for the filled unsaturated polyester resin (150 and 200 phr, respectively). In these plots one can again see the secondary transition centered around $-100°C$ as a max-

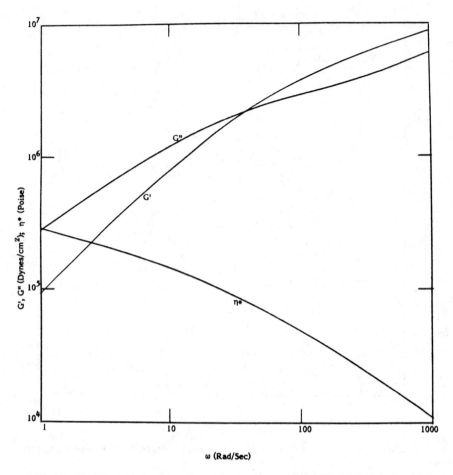

FIG. 14—G', G'', and η^* versus ω for calcium carbonate filled (200 phr) thickened polyester resin.

imum in G'' and tan δ. The polyester softens above 100°C and shows a maximum tan δ at approximately 200°C. The overall properties are observed to be much higher than with the polyester resin alone, and one can see an incremental increase in moduli over the whole range for the 200 phr sample versus the 150 phr sample. The most dramatic changes over the unfilled sample are the increases in modulus and the temperature of the maximum in the damping curve.

Reinforced Filled Polyester Compound

In Figs. 20 and 21, G', G'', and tan δ are plotted versus T for the 10 and 20 percent glass filled materials, respectively. In these plots one can again see

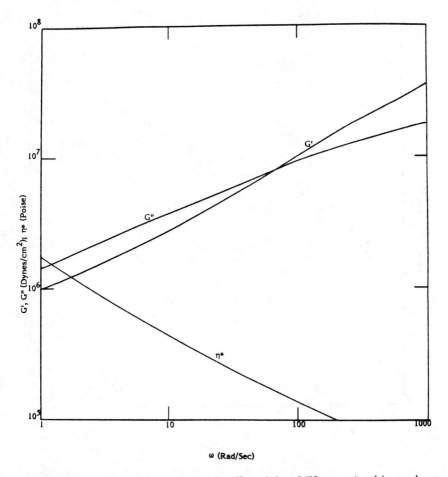

FIG. 15—*G'*, *G"*, *and η* versus ω for glass fiber reinforced (10 percent), calcium carbonate filled (150 phr) thickened polyester resin.*

the secondary transition at $-100°C$ as evidenced by the maxima in tan δ and *G"*. The properties of the materials have risen over those of the filled materials, but not as significantly as would be expected from comparisons to tensile or flexural modulus data. This is to be expected, since dynamic shear modulus data are more dependent on the polyester matrix than on the reinforcement [4].

In Fig. 22, the changes in *G'* associated with compounding steps are demonstrated. This plot shows *G'* for the non–cross-linked alkyd polyester, the cross-linked polyester resin, the filled cross-linked material, and the filled reinforced molding compound. One can see a progressive increase in the modulus as a function of each additional compounding step. One can also

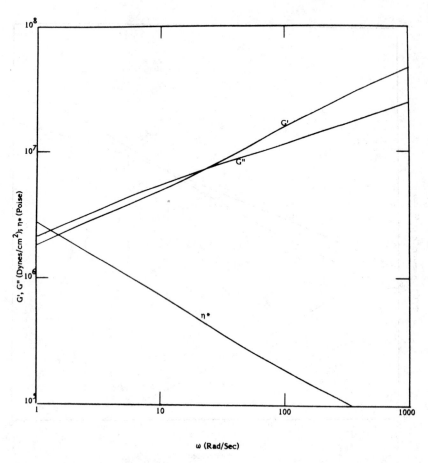

FIG. 16—G', G", and η* versus ω for glass fiber reinforced (20 percent), calcium carbonate filled (150 phr) thickened polyester resin.

see the increase in the temperature stability of the filled and the glass reinforced specimens. The decline in high temperature modulus is much less severe for the glass filled material than for any of the other materials.

Conclusions

It has been demonstrated that SMC and BMC materials undergo substantial changes from initial manufacture through the molding cycle. These changes make the behavior of SMC and BMC very complex; a technique such as dynamic mechanical testing is necessary to demonstrate the role of various constituents in the total formulation.

In this investigation, it was demonstrated that chemical thickeners impart

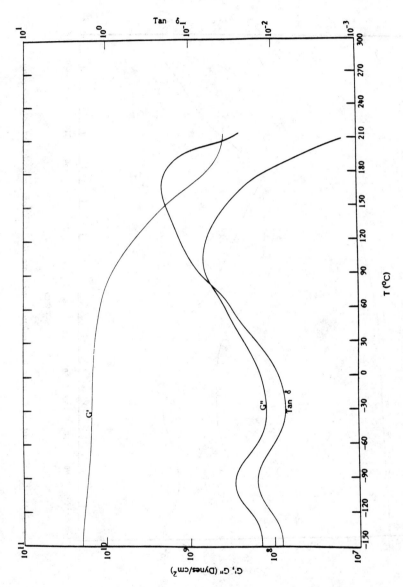

FIG. 17—*G', G", and tan δ versus T for styrene cross-linked polyester resin.*

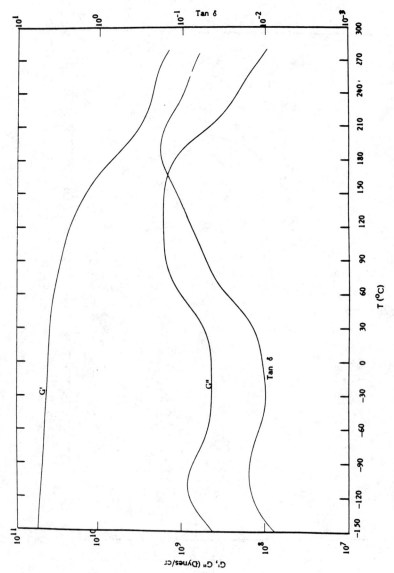

FIG. 18—G', G", and tan δ versus T for calcium carbonate filled (150 phr) cross-linked polyester resin.

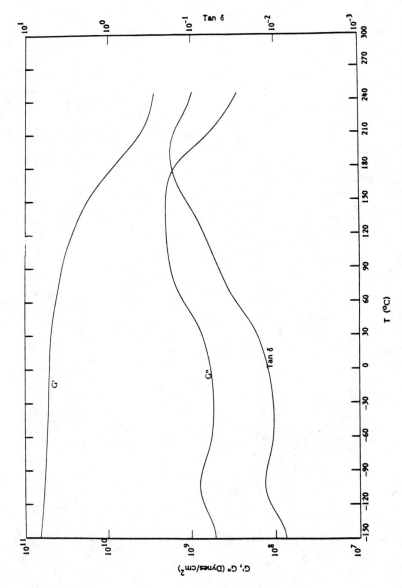

FIG. 19—G', G", and tan δ versus T for calcium carbonate filled (200 phr) cross-linked polyester resin.

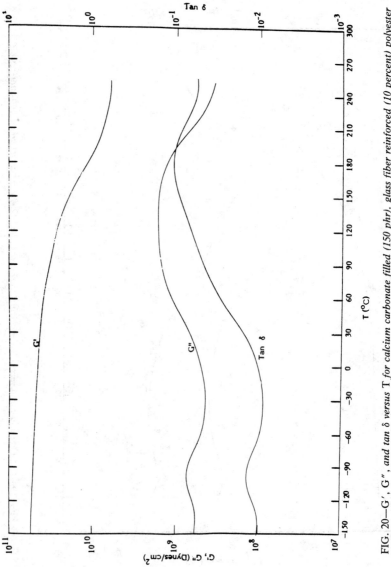

FIG. 20—G', G", and tan δ versus T for calcium carbonate filled (150 phr), glass fiber reinforced (10 percent) polyester resin.

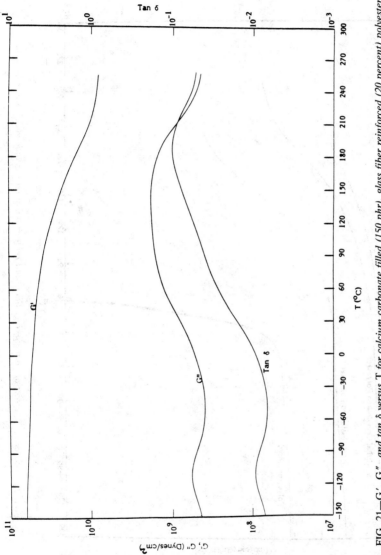

FIG. 21—G', G", and tan δ versus T for calcium carbonate filled (150 phr), glass fiber reinforced (20 percent) polyester resin.

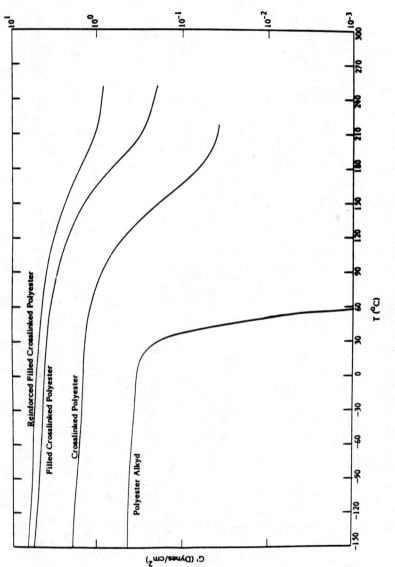

FIG. 22—G' versus T: effect of cross-linking, filler, and reinforcement on dynamic elastic shear modulus.

elasticity to the compound due to an apparent increase in the molecular weight of the polyester resin. This effect is necessary to provide sufficient elasticity to prevent glass separation during mold flow. Filler materials were demonstrated to raise the elastic and viscous shear moduli and complex viscosity of the compound in both the fluid and the cross-linked state. Fiber reinforcement addition was seen to further increase the modulus of both the melt and the cross-linked material, while imparting a yield behavior to the non–cross-linked material.

Information from this type of investigation, when added to the information collected from standard test methods, yields a significant increase in the understanding of the role of the various additives of SMC and BMC compositions. This type of testing leads to an overall understanding of the various stages of material before it is processed and molded into the final part.

References

[1] Burns, R., Lynskey, B. M., and Gandhi, K. S., "Variability in Sheet Molding Compound," in *Research Projects in Reinforced Plastics,* The Plastics and Rubber Institute, Reinforced Plastics Group (RPG), Fourth Conference, 1976.

[2] "Thermal Mechanical Spectrometer Viscoelastic Properties from Fluid to Solid Composites," product literature, Rheometrics, Inc., Union, N. J.

[3] Rouse, P. E., *Journal of Chemical Physics,* Vol. 21, 1953, p. 1272.

[4] Read, B. E. and Dean, G. D., *The Determination of Dynamic Properties of Polymers and Composites,* Wiley, New York, 1978, p. 19.

[5] Rapp, R. S., "Sheet Molding Compounds: The Effects of Isophthalic Polyester Processing," in *Proceedings,* 34th Annual SPI Conference, Section 16-F, Reinforced Plastics/Composites, 1979, pp. 1-6.

[6] Ludwig, C. and Collister, J. in *Proceedings,* 34th Annual SPI Conference, Section 24-C, Reinforced Plastics/Composites, 1979, p. 1.

[7] Soong, D., Shen, M., and Hong, S. D., *Journal of Rheology,* Vol. 23, No. 3, pp. 301-322.

[8] Gruskiewicz, M. and Collister, J., "Analysis of the Thickening Reaction of an SMC Resin Through the Use of Dynamic Mechanical Testing," in *Proceedings,* 35th Annual SPI Conference, Section 7-E, Reinforced Plastics/Composites, 1980, pp. 1-9.

R. M. Alexander,[1] R. A. Schapery,[1] K. L. Jerina,[1] and B. A. Sanders[2]

Fracture Characterization of a Random Fiber Composite Material

REFERENCE: Alexander, R. M., Schapery, R. A., Jerina, K. L., and Sanders, B. A., "Fracture Characterization of a Random Fiber Composite Material," *Short Fiber Reinforced Composite Materials, ASTM STP 772,* B. A. Sanders, Ed., American Society for Testing and Materials, 1982, pp. 208–224.

ABSTRACT: A short-glass fiber composite material, SMC-R50, was tested under uniaxial loading at ambient temperature and humidity conditions over two orders of magnitude of loading rate. The test method, experimental results, and analytical characterization of fracture behavior are described. Baseline information on the strength of 1.27-cm and 2.54-cm wide unnotched tension specimens indicate that effects of size and test orientation (relative to molding orientation) are not significant. Fracture behavior of notched samples was studied on specimens of two different widths (2.54 and 5.08 cm); each sample had two symmetrically cut, crack-like notches. For each sample size, the effect of four different notch lengths and loading rates on ultimate strength, ultimate strain, and the overall stress-strain curve were investigated. Loading rate was found to have no significant effect on sample response. The stress-strain curves of notched and unnotched samples in terms of nominal stress σ and strain ϵ obey the power law $\sigma \sim \epsilon^{0.78}$ out to failure. In spite of this nonlinearity, it was possible to characterize the fracture stress behavior using linear elastic fracture mechanics theory and a constant fracture toughness. The effect of notch length on sample compliance and on failure strain was found to obey this theory if $\epsilon^{0.78}$ is used in place of ϵ.

KEY WORDS: composite material, glass fibers, fracture (materials), failure, fracture tests, toughness, tension tests

The strength of a monolithic or composite material is dependent on the size of preexisting flaws and the resistance of the material to flaw growth. For a linear elastic material, a basic material parameter that characterizes this resistance is the so-called fracture toughness [1].[3] This property is very im-

[1]Texas A&M University, College Station, Tex., 77843.
[2]Plastic Materials Characterization Department, General Motors Technical Center, Warren, Mich. 48090.
[3]The italic numbers in brackets refer to the list of references appended to this paper.

portant for many engineering considerations. For example, one can specify the largest natural or artifically created flaw or sharp notch that can be tolerated by a given structure under a given set of loads. It therefore provides a criterion for acceptance or rejection of a structural member based on the results of nondestructive inspection.

In this paper we show that linear elastic fracture mechanics principles can be employed to characterize the toughness and stiffness behavior of a notched, random fiber composite SMC-R50 at room temperature and humidity. The stress-strain curve is found to be quite nonlinear, which would seem to indicate that the more general J-integral method should be used [2]. However, simply by using a nonlinear measure of sample deformation in place of deformation itself, this linear theory is found to accurately characterize and predict behavior. (In a preliminary investigation of the use of the J-integral, we obtained essentially the same results as from linear elastic fracture mechanics, but this matter requires further study.)

Other investigators have applied different versions of linear elastic fracture mechanics theory to random composites. Three such studies are in Refs 3 to 5.

Gaggar and Broutman [3] conducted an investigation to determine the applicability of linear elastic fracture mechanics to a randomly oriented discontinuous fiber epoxy composite. The candidate stress intensity factor K_Q was determined using single edge notched, double edge notched, and notched bend tests. Load–crack mouth opening displacement curves were used to determine K_Q from the procedures of ASTM Test for Plane-Strain Fracture Toughness of Metallic Materials (E 399). The compliance method was also used to determine K_Q to check the validity of using the K-equations for isotropic materials. The investigators found that the values of K_Q from the compliance method were about 20 percent lower than the values obtained from load-displacement curves. The effect on K_Q of specimen thickness, notch-root radius, specimen width, and fiber volume fraction were investigated. Fracture toughness mechanisms from an energy viewpoint were discussed and it was concluded that fracture energy is dominated by the fiber-matrix debonding energy and the fiber pull-out energy. In another paper, Gaggar and Broutman [4] investigated the crack growth resistance of random glass fiber composites. Crack growth resistance curves (R-curves) were developed for epoxy and polyester composites to examine the relationship between crack growth resistance and crack extension. Single edge notch tensile specimens with various crack lengths were used. A candidate stress intensity factor was determined for each material from the R-curve for that material. The investigators then used the R-curves to predict the fracture strength of a plate containing a central hole subjected to tensile loading. Predicted and measured failure stresses compared favorably. Their results suggest that the R-curve concept is a viable approach to the study of crack growth in random fiber composites. Owen and Bishop [5] tested double edge notched tension specimens made from various forms of glass reinforced polyester resin. They determined critical stress intensity factors K_c from load—crack-opening displacement data using linear elastic fracture mechanics and stress intensity factor expressions for isotropic materials. They found that K_c increased with crack length for chopped strand mat composites, but a correction to the crack length based on equivalent yield stress produced an independence of K_c with crack length. This K_c value was then used, with reasonable success, to predict the failure stresses of plate specimens with a central hole.

In the next section the experimental procedure for the present investigation is briefly stated. Results concerning specimen size and orientation effects on unnotched samples are then given. Also, strain gage and extensometer data are compared in order to check the two methods of following sample deformation; we were concerned that sample inhomogeneity may lead to erroneous results using strain gages.

The last section describes the fracture mechanics theory employed to characterize fracture behavior and predict the effect of notches on overall sample compliance and strain. Also, experimental and theoretical results are compared.

Information on creep characteristics of the same material studied here is given in Ref 6.

Experimental Procedure

Fracture toughness tests were performed on an MTS servocontrolled hydraulic testing system in the load-control mode. A double edge notch tension specimen was used for all tests. Each specimen was pulled to failure at the desired loading rate while recording load, far-field displacement, and crack opening displacement. For each test, load versus far-field displacement (extensometer) and load versus crack-opening displacement (clip gage) were recorded on a dual-channel X-Y plotter. In addition, load versus time was recorded to monitor load rate. Crack-opening displacement was measured to indicate the onset of unstable crack growth. A schematic of the test specimen indicating the parameters recorded is shown in Fig. 1.

Results

An investigation of baseline fracture behavior was conducted to determine the effect of specimen size and orientation on ultimate tensile strength. Full scale and half scale unnotched specimens were failed under the same conditions. Tables 1 and 2 summarize the ultimate strength for the various sizes. The Weibull distribution was used to analyze the data. The shape parameter is a measure of the dispersion and the scale parameter is a measure of the central tendency. The large specimens were of the Illinois Institute of

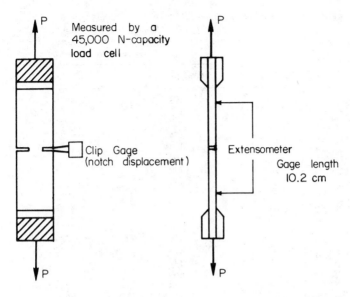

FIG. 1—*Double edge notch specimen with measured parameters indicated.*

Technology Research Institute (IITRI) design and the small specimens were scaled to one half. These tests indicated that the scatter in strength of unnotched samples is the same order of magnitude as any variation due to orientation or specimen size.

The double edge notch (DEN) specimen geometry is shown in Fig. 2. Table 3 summarizes the test conditions and gives the specific geometric values used. Three replicate tests were done at each condition for the 2.54-cm-wide specimens while one test at each condition was conducted for the 5.08-cm-wide specimens.

Unnotched tension specimens were tested to obtain basic material behavior. Figure 3 shows typical stress-strain curves for three different loading rates. Strain was measured with an extensometer over a gage length of 102 mm. Note that the rate effect is small for the rates considered. It was found that the stress-strain behavior for the notched as well as the unnotched specimens could be fit to a power law *out to failure*, as illustrated in Fig. 4:

$$\sigma = C\epsilon^p \tag{1}$$

A limited number of tests were conducted with centrally mounted strain gages used to measure strain. Strain gage data for two loading rates are shown in Figs. 4 and 5, with extensometer data shown for comparison. It is seen that the localized strain behavior is very close to that measured with the extensometer.

TABLE 1—*Effect of specimen width and orientation on the ultimate strength (MPa) of Panel 5-64.*

Width	Orientation	
	Longitudinal	Transverse
2.54 cm	153	157[a]
	167	165[a]
	168[a]	172
	174[a]	172
	163	...
	165	...
	179	...
	167	...
	161	...
1.27 cm	157[a]	125
	161	116[a]
	170	137[a]
	173	133
	147	134
	160	130[a]
	144	123
	157[a]	124
	141[a]	161[a]
	145[a]	154
	160[a]	147[a]
	169[a]	154
	160[a]	143
	160	150[a]
	...	165[a]
	...	161[a]

[a]Failure in tab region.

TABLE 2—*Weibull distribution parameters for ultimate strength of Panel 5-64.*

Width, cm	Orientation[a]	No. of Tests	Shape Parameter	Scale Parameter, MPa
2.54	L	4	18	170
2.54	L	5	21	171
2.54	T	4	23	170
1.27	L	6	18	165
1.27	L	8	16	159
1.27	T	8	19	131
1.27	T	8	21	158
2.54	L&T	13	26	170
1.27	L&T	30	11	156
2.54 and 1.27	T	20	9	154
2.54 and 1.27	L	23	18	165

[a]L = longitudinal; T = transverse.

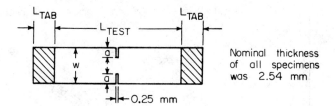

FIG. 2—*Double edge notch specimen geometry.*

TABLE 3—*Test conditions and DEN specimen geometry.*[a]

Load Rate, N/s	a, mm	w, cm	L_{TAB}, cm	L_{TEST}, cm
45,200,965,4450	0, 1.9, 3.8, 5.7, 7.6	2.54	3.8	15.24
89,400,1850,8900	3.8, 7.6, 11.4, 15.2	5.08	5.08	22.86

[a]See Fig. 2 for definitions of L_{TAB} and L_{TEST}.

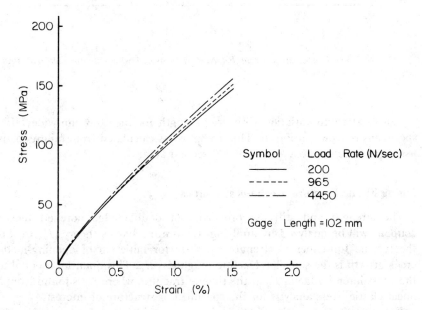

FIG. 3—*Stress-strain curves based on load-displacement data for 25.4-mm-wide unnotched tension specimens.*

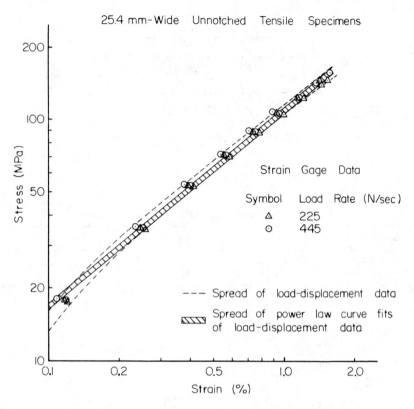

FIG. 4—*Comparison of stress-strain behavior as measured with extensometer and strain gages—logarithmic scale.*

Tensile strength variation with crack length for the 50.8-mm-wide DEN specimens is shown in Fig. 6. The strength was calculated from failure loads using gross section area as well as net section area.

Linear Elastic Fracture Mechanics Analysis

The strength and stiffness properties of double edge notched tensile coupons will be analyzed by applying fracture mechanics theory [*1*]. In this theory, the fundamental strength property for initiation of opening-mode crack growth is the so-called fracture toughness K_c. This quantity is equal to the stress-intensity factor K_I at the time of fracture, where K_I is found from a linear elastic stress analysis for the specimen or structure of interest.

For a relatively long ($2w/L \leq 3$ where w = width and L = gage length) tensile coupon with sharp double edge notches (as in Fig. 2), the stress intensity factor is [*1*]:

$$K_I = Y\sigma\sqrt{a} \tag{2}$$

where

$$Y \equiv 1.122\sqrt{\pi}\,f \tag{3}$$

and f is a finite-width correction factor; this factor is a function of the crack length-to-specimen width ratio a/w, and $f \rightarrow 1$ when $a/w \rightarrow 0$. The factor $f = f(a/w)$ is given in Ref 7, and is described accurately by the function

$$f \simeq \left(1 + 0.122\cos^4\left(\frac{\pi a}{w}\right)\right)\sqrt{\frac{w}{\pi a}\tan\frac{\pi a}{w}}\ \bigg/ 1.122 \tag{4}$$

for $0 \leq 2a/w < 1$.

Denote the stress at fracture by σ_f. Then, from Eq 2, with $K_I = K_c$,

$$K_c = Y\sigma_f\sqrt{a} \tag{5}$$

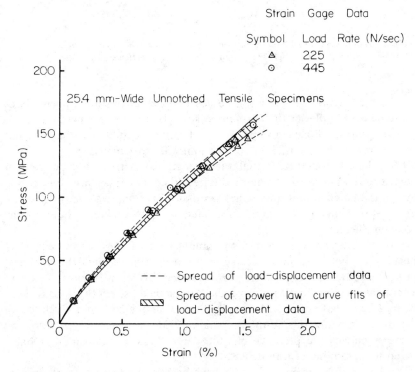

FIG. 5—*Comparison of stress-strain behavior as measured with extensometer and strain gages.*

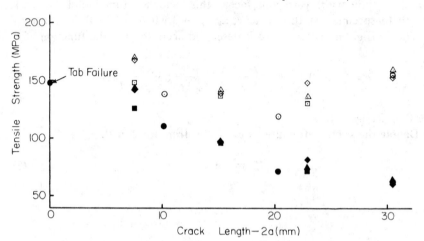

FIG. 6—*Tensile strength variation with crack length (open symbols denote strength based on net section area; solid symbols denote strength based on gross section area).*

Thus, if the fracture behavior can be described by a constant value of fracture toughness, a plot of the normalized strength $Y\sigma_f$ against notch length a on log-log paper will have a slope of $-\frac{1}{2}$. Figures 7 and 8 show the data and linear regression lines for 25.4 and 50.8 mm wide specimens.

It is inferred from Fig. 7 that the strength is relatively independent of notch length when $a \lesssim 2.5$ mm; this result implies there are pre-existing flaws in the notched and unnotched samples whose "equivalent" edge-crack size is approximately 2.5 mm; this value is approximately equal to the specimen thickness.

It is of interest to consider one implication of this equivalent size. Specifically, the stress intensity factor for a central through-the-thickness crack of length $2a_p$ is $K_I = \sqrt{\pi a_p} \ \sigma$. Equating this stress intensity factor to the critical value in Eq 5 and using the "unnotched" strength σ_f for σ yields $2a_p \simeq 2.5a \simeq 6.3$ mm, which is more than twice the specimen thickness of 2.54 mm. Although this result is for a very idealized crack geometry, we believe it implies the presence of quite large flaws or soft zones probably related to a grouping of bundle ends.

The values of $Y\sigma$ in Figs. 7 and 8 have been multiplied by $\sqrt{a}$ in order to obtain the fracture toughness; the result is plotted in Figs. 9 and 10. No

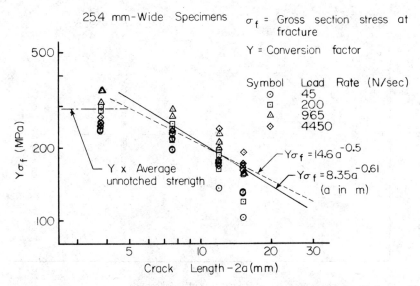

FIG. 7—*Variation of normalized fracture stress with crack length for the 25.4-mm-wide specimens. (The solid line is a least squares fit excluding the 2a = 3.81 mm data and the 45 N/s load rate data. The dashed line shows a fit of the same data using linear fracture mechanics.)*

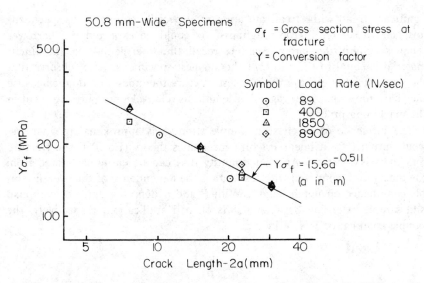

FIG. 8—*Variation of normalized fracture stress with crack length for the 50.8-mm-wide specimens. (The line is a least squares fit of the data.)*

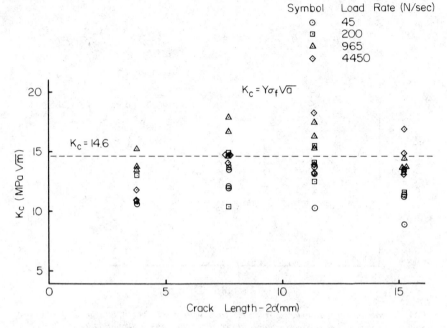

FIG. 9—*Critical stress intensity factor for the 25.4-mm-wide specimens.*

significant trend with stress rate appears, especially compared to specimen-to-specimen scatter in the toughness. As could be expected the narrower samples exhibit the largest scatter; recall the fiber length in the uncut material is equal to the width of the smallest specimens. Also, observe that the smallest samples have the smallest fracture toughness, no doubt because they have the smallest average fiber length, as relatively more fibers are cut in the machining process.

The influence of notches on sample strength is approximately what one would predict from linear fracture mechanics theory (Eq 5) if the fracture toughness is constant. Another quantity that can be easily calculated from the theory and compared with test data is the dependence of the overall sample compliance on notch length. With P and u denoting applied axial load and sample extension between points of load application, respectively, the sample compliance S is defined as

$$S \equiv \frac{u}{P} \tag{6}$$

For a linear elastic material, S is independent of load but is a function of notch length a.

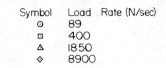

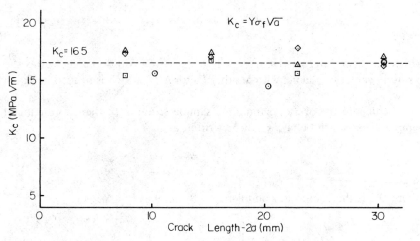

FIG. 10—*Critical stress intensity factor for the 50.8-mm-wide specimens.*

The compliance will be related to K_I through energy considerations [1]. It is helpful to introduce an expression for the strain energy, V, in one half the sample width:

$$V = \frac{1}{4} Pu = \frac{1}{4} \frac{u^2}{S} \qquad (7)$$

For an elastic material with fixed grips, the strain energy released per unit area of crack growth Bda, where B is sample thickness, is

$$-\frac{\partial V}{B \partial a} = \frac{u^2}{4S^2 B} \frac{\partial S}{\partial a} = \frac{P^2}{4B} \frac{\partial S}{\partial a} \qquad (8)$$

In turn, this release rate is equal to K_I^2/E_c, where E_c is the Young's modulus for the composite. Thus, using Eq 3, we obtain

$$\frac{\partial S}{\partial a} = \frac{4BK_I^2}{P^2 E_c} = \frac{15.76 \, a f^2}{B E_c w^2} \qquad (9)$$

Then, by integrating and using the fact that the sample compliance without notches is $L/(BwE_c)$, where L is the gage length, we find

$$S = \frac{L}{BwE_c}\left(1 + \frac{7.88a^2}{wL}g\right) \tag{10}$$

where

$$g = 1 + 2\frac{w^2}{a^2}\int_0^{a/w}\rho(f^2 - 1)\,d\rho \tag{11}$$

which has been integrated numerically using f from Eq 4. Table 4 gives the resulting values of g along with another factor g/f^4, which is needed in Eq 16 below.

Let ϵ denote the overall or nominal sample strain u/L; then, the nominal sample stress-strain behavior can be written as

$$\epsilon = D_{cn}\sigma \tag{12}$$

where D_{cn} is the composite compliance with notches:

$$D_{cn} = \frac{1}{E_c}\left(1 + \frac{7.88a^2}{wL}g\right) \tag{13}$$

Generalization of Sample Compliance for Nonlinear Behavior

As reported above, all of the nominal stress-strain behavior of notched and unnotched samples is accurately characterized from small strains out to failure by the power law

$$\sigma = C\epsilon^p \tag{14}$$

where p is constant but C is a function of notch length. Inasmuch as C is independent of strain, we shall assume its dependence on notch size is that found from linear elastic theory (Eq 13); hence

$$\frac{1}{C} = \frac{1}{E_c}\left(1 + \frac{7.88a^2}{wL}g\right) \tag{15}$$

In order to check this result, we shall use Eqs 14 and 15 together with the failure stress (Eq 5), in order to predict the locus of failure points in Fig. 11. These equations serve to predict the nominal failure strain ϵ_f as a function of failure stress σ_f:

$$\epsilon_f{}^p = \frac{\sigma_f}{E_c}\left(1 + 0.51\frac{K_c{}^4 g}{wL\sigma_f{}^4 f^4}\right) \tag{16}$$

TABLE 4—*Values of* g *and* g/f⁴.

a/w	g	g/f^4
Up to 0.15	1.0	1.0
0.20	1.0	0.967
0.25	1.015	0.880
0.30	1.050	0.736
0.35	1.117	0.544
0.40	1.244	0.327

The predicted failure locus, $\sigma_f - \epsilon_f$, for both specimen widths is shown in Fig. 11. Equation 16 was used together with the following constants found by averaging overall data: $K_c = 14.6$ and 16.5 MPa$\sqrt{m}$ for the 25.4 and 50.8 mm wide specimens, respectively (cf. Figs. 9 and 10). Also, $E_c = 110$ MPa (with ϵ_f in percent) and $p = 0.78$. (The stress-strain behavior of each unnotched specimen was fit to a power law, $\sigma = E_c \epsilon^p$. The E_c values were then averaged geometrically and the p values arithmetically to obtain these values.) Curves A and C in Fig. 11 were obtained from Eq 16, but the width correction was neglected; that is, $f = g = 1$ was used. This correction is significant only in the low stress level range, but it serves to bring the theory into agreement over the entire stress range. The prediction with f and g included is shown in Fig. 11 by Curves B and D.

The fact that Eqs 5 and 16 serve to accurately characterize the effect of sharp notches on specimen compliance and failure behavior means in effect that linear elastic fracture mechanics applies. In order to use this linear theory, however, it is necessary to replace ϵ by ϵ^p in predicting overall sample strain.

As one final matter of interest, we have drawn in Fig. 11 the linear regression line for failure of 25.4-mm-wide unnotched specimens. The slope of this line is practically the same as predicted by the micromechanics theory [6] using data obtained from high-temperature low-stress creep tests for a composite in which the matrix modulus is zero; that is, the limiting modulus in Ref 6 is approximately 7.2 GPa. Stress-strain behavior at high stresses therefore can be predicted by assuming the total force is that carried by a linear elastic fiber network plus a constant. This constant in Fig. 11 is 40 MPa, which is practically equal to the tensile strength of 50 MPa for matrix-only samples [6]. We have estimated the residual stress at room temperature to be about 10 MPa, which serves to help confirm this interpretation of specimen behavior at high stresses.

Conclusions

The composite material SMC-R50 is quite nonlinear; however, it exhibits the same overall deformation response as a linear material with or without

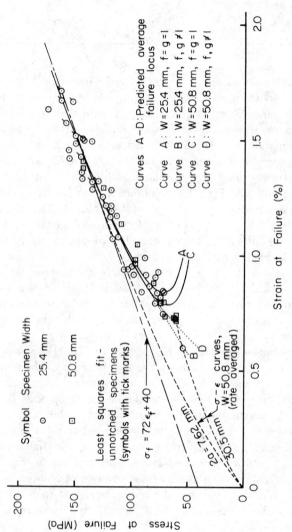

FIG. 11—Measured and predicted stress and strain at failure.

notches if a nonlinear measure of average strain ϵ^p is used in the linear theory in place of strain. This observation could prove useful in design analysis, but further work is needed to establish the extent of such simplicity.

The fracture data for unnotched samples fall along a straight line of stress versus strain that can be predicted by assuming the material consists of an elastic glass fiber network that is embedded in a perfectly plastic material with a yield stress equal to the strength of the matrix. This behavior is analogous to that often found for laminates consisting of unidirectional plies when one or more plies are taken above their ultimate strain as obtained from a unidirectional ply test. The slope of this strength-ultimate strain line is also predicted from the low stress level creep data and micromechanics theory in Ref 6.

There is a large amount of scatter in failure stress and strain for both notched and unnotched samples. However, the data tend to scatter along the theoretically predicted lines of σ_f versus ϵ_f. Therefore this scatter is believed to be due to randomness in dominant flaw size and its interaction with the local effect of the short fiber length, rather than to mechanical properties. Also, sample width has a significant effect in that the 25.4-mm-wide samples exhibited much more scatter and a smaller fracture toughness than the 50.8-mm samples. The authors plan to test larger samples to determine if 50.8-mm samples are large enough to provide a reliable value of fracture toughness.

Fracture toughness of the wide samples was found to be essentially constant with a value of $K_c = 16.5$ MPa$\sqrt{m}$ for all crack lengths and loading rates, where the rates varied by a factor of 100. This same toughness can be used to predict the strength of unnotched samples by assuming these samples contain pre-existing "natural" edge notches of length approximately equal to specimen thickness. We have found this point of view to be more consistent with our data than one based on Whitney and Nuismer's stress criteria [8]; however, this interpretation is very tentative since the data are quite limited.

The loading rate insensitivity is not surprising in view of the weak time-dependence of creep compliance reported in Ref 6. A detailed study on the effect of viscoelastic properties of the matrix on strength would have to include a significantly extended range of rates, including the limiting case of creep-rupture. Changes in temperature may produce effects analogous to those in loading rate, but the creep data in Ref 6 seem to indicate a complete analogy does not exist, as the resin is not thermo-rheologically simple.

Acknowledgments

The authors gratefully acknowledge the financial support of this research by the Plastics Materials Characterization Department, Manufacturing Development Division, General Motors Corporation. The capable assistance of Mr. C. Fredericksen in conducting the experimental program is greatly appreciated. Mr. M. McEndree and Mr. M. Kerstetter prepared the samples from the compression molded sheets of SMC-R50 supplied by the General Motors Corporation.

References

[1] Hertzberg, R. W., *Deformation and Fracture Mechanics of Engineering Materials*, Wiley, New York, 1976.
[2] Rolfe, S. T. and Barsom, J. M., *Fracture and Fatigue Control in Structures*, Prentice-Hall, Englewood Cliffs, N.J., 1977.
[3] Gaggar, S. K. and Broutman, L. J. in *Flaw Growth and Fracture*, ASTM STP 631, American Society for Testing and Materials, 1977, pp. 310–330.
[4] Gaggar, S. K. and Broutman, L. J., *Journal of Composite Materials*, Vol. 9, 1975, p. 216.
[5] Owen, M. J. and Bishop, P. T., *Journal of Composite Materials*, Vol. 7, April 1973, p. 146.
[6] Jerina, K. L., Schapery, R. A., Tung, R. W., and Sanders, B. A., this publication, pp. 225–250.
[7] Tada, H., Paris, P. C., and Irwin, G. R., *The Stress Analysis of Cracks Handbook*, Del Research Corp., Hellertown, Pa., 1973, pp. 2.6–2.7.
[8] Nuismer, R. J. and Whitney, J. M. in *Fracture Mechanics of Composites*, ASTM STP 593, American Society for Testing and Materials, 1975, pp. 117–142.

K. L. Jerina,[1] *R. A. Schapery,*[1] *R. W. Tung,*[2] *and*
B. A. Sanders[2]

Viscoelastic Characterization of a Random Fiber Composite Material Employing Micromechanics

REFERENCE: Jerina, K. L., Schapery, R. A., Tung, R. W., and Sanders, B. A., **"Viscoelastic Characterization of a Random Fiber Composite Material Employing Micromechanics,"** *Short Fiber Reinforced Composite Materials, ASTM STP 772,* B. A. Sanders, Ed., American Society for Testing and Materials, 1982, pp. 225-250.

ABSTRACT: The mechanical creep and recovery behavior for uniaxial loading of a short glass-fiber reinforced polyester composite, SMC-R50, with random fiber orientation was investigated experimentally and theoretically. The strain at different stress levels and temperatures was measured during isothermal tests involving several loading-unloading cycles; each cycle consisted of a period of constant load and zero load. Repeatable creep and recovery response was observed only after several cycles; an accelerated mechanical conditioning method was used, in which the duration of each cycle was progressively increased. It is shown that the thermally complex, linear viscoelastic response of SMC-R50 in the mechanically conditioned state can be described by the time-dependent and temperature-dependent behavior of the matrix system and micromechanics calculations of the effective properties of the composite. The data analysis methods for parameters of the model are discussed and illustrated. Master curves for the matrix and for the *in situ* matrix, the matrix in the composite, were obtained and found to be virtually the same. In order to achieve this agreement it was necessary to use a micromechanics analysis in which the reinforcement is properly modeled as viscoelastic, orthotropic ribbons rather than isotropic, elastic fibers.

KEY WORDS: composite material, random fiber composite, sheet molding compound, viscoelasticity, creep and recovery, micromechanics

Fiber-reinforced polymers are now being used extensively in aerospace applications where materials with high strength, high stiffness, and low density are required. For some time, the automotive industry has applied polymers

[1]Mechanics and Materials Research Center, Texas A&M University, College Station, Tex. 77843.
[2]General Motors Manufacturing Development, General Motors Technical Center, Warren, Mich. 48090.

as unfilled, particle-filled, and fiber reinforced systems. As a primary means of reducing automobile fuel consumption, considerable effort is now directed to greater structural use of composites to reduce vehicle weight.

Where long-term strength and dimensional stability are needed, a thorough understanding of mechanical and failure behavior is essential. In contrast to common structural metals, the response of polymers in automobile service environments may vary significantly with time, loading frequency, temperature, humidity, and exposure to various chemicals, in which high stresses affect sensitivity of response to these parameters [1-7].[3] Although many analytical and experimental studies have been conducted, we are now only beginning to understand the behavior of random-fiber automotive composites and to develop the analytical tools and database essential to achieving efficient, economical, and reliable structural designs.

The type of composite material studied herein is a so-called sheet molding compound (SMC), which is used by the automobile industry and other industries where parts must be mass-produced at low cost. The specific composite tested is SMC-R50; it consists of a polyester matrix filled with calcium carbonate particles and reinforced with 50 weight percent of 25.4-mm (1-in.)-long randomly oriented glass fibers.

The mechanical response of plastics under uniaxial stress is commonly characterized in terms of a creep compliance D:

$$D \equiv \epsilon(t)/\sigma \tag{1}$$

where stress σ (referred to initial area) is a constant applied at $t = 0$ and strain ϵ is a function of time. The ability of the material to recover or return to its initial length is conveniently represented by the recovery compliance D_r:

$$D_r \equiv \epsilon_r(t)/\sigma \tag{2}$$

where $\epsilon_r(t)$ is the strain after a constant stress σ is removed.

In general, these compliances depend on stress and temperature, as well as other physical and chemical parameters such as moisture content. A composite typically contains many flaws such as microcracks, crazes, and fiber-matrix disbonds. Creep strain of SMC material during the first load application often exhibits sudden jumps [1-3], probably due to sudden microflaw growth and subsequent arrest.

This erratic behavior, as well as more regular creep straining due to flaw growth, such as is exhibited by continuous glass and graphite fiber-reinforced plastics [8,9] in off-angle tests, greatly complicates the material characterization task. On the other hand, if one desires to determine the basic creep characteristics of the material, rather than first-cycle creep response, the problem can be greatly simplified by using mechanically condi-

---

[3]The italic numbers in brackets refer to the list of references appended to this paper.

tioned samples. So long as the stress level is not too high, a nearly constant flaw state is reached after several load-unload cycles at a given stress level; this condition is revealed by a creep compliance that does not change noticeably with a limited amount of further cycling. Creep and recovery response was characterized in Refs 6, 8, and 9 by using mechanically conditioned specimens.

All creep and recovery data given in this paper are for the mechanically conditioned state, with the conditioning history as shown in Fig. 1. The use of progressively increasing cycle periods, instead of constant periods, reduces the time required to perform this conditioning. In all of the analytical characterization work, we employed only data at stress levels equal to or less than the conditioning stress and for periods t' (compare Fig. 1b) that did not exceed twice that of the last conditioning cycle.

Figures 2 to 4 show representative data for the logarithm of creep compliance versus the logarithm of time at different stresses and temperature. In contrast to the nonlinear behavior of other composites [5,6], the slope decreases with increasing stress (compare Figs. 3 and 4). Also, considering these and other test results, it is not possible to form a master curve of creep compliance (a single curve consisting of data for various stresses and/or temperatures) with respect to either stress level or temperature; it is usually possible to construct these master curves of D or log D through horizontal and vertical shifting of the data [10]. Thus prediction of long-time response from short-time data cannot be accomplished by using standard techniques that rely on the ability to form a master property curve for the composite. This complex behavior is believed to be due in part to the fiber domination of the composite stiffness. Indeed, we will show that the *in situ* resin compliances at different temperatures can be used to develop a master curve in the linear viscoelastic range of behavior; this master curve together with micromechanics provides the means to predict long-time behavior from short-time tests. Presumably for high stresses the nonlinear viscoelastic characterization technique employed by Lou and Schapery [6] is applicable if average stress in the resin is used instead of applied stress, but our nonlinear data have not yet been analyzed in this manner.

In the next two sections we use micromechanics theory to develop the relation between matrix creep compliance and that for a composite with viscoelastic ribbon reinforcement. The remaining sections are concerned with the experimental work and data analysis based on this relation.

Fracture data and analysis for notched and unnotched specimens of the same material investigated here are given in Ref 11.

Micromechanics Theory for Linear Elastic Behavior

The Halpin-Tsai micromechanics equations for fibrous composites [12] will now be used with lamination theory to relate the composite's effective

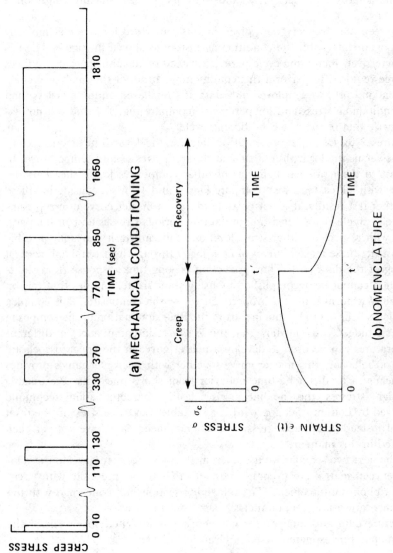

FIG. 1—*Mechanical conditioning history (a) as a sequence of creep-recovery periods (b).*

SYMBOL	TEMP(C)	STRESS(PA)	T PRIME(SEC)	CYCLE #	SPEC #
□	23.9	1.66200E+07	640	7	B16
☉	40.6	1.66200E+07	640	7	B16
▲	57.2	1.64900E+07	640	7	B16
+	76.6	1.66200E+07	640	7	B16
×	93.3	1.66200E+07	640	7	B16

LOG10 CREEP COMPLIANCE(1/PA) VERSUS LOG10 TIME(SEC)

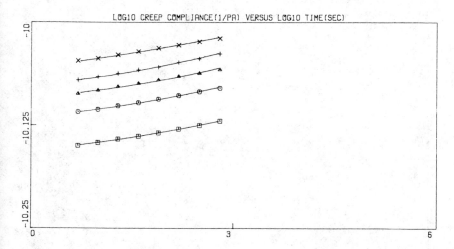

FIG. 2—*Creep compliance at a creep stress of 10 percent of the room-temperature ultimate strength and at 23.9, 40.6, 57.2, 76.6, and 93.3°C.*

Young's modulus E_c and Poisson's ratio ν_c to elastic constituent properties and fiber volume fraction. The next section extends the results to viscoelastic behavior. As will be shown from a study of the experimental data, the fibers in this theoretical model cannot be assumed to have isotropic or elastic behavior. Rather, they must be treated as transversely isotropic, viscoelastic material. The physical basis for this generalization is that the individual glass fibers in SMC material are actually grouped into ribbon-shaped bundles consisting of approximately 200 fibers during processing. These bundles remain as such in the molded composite, as can be seen from Fig. 5. The bundles are approximately rectangular with a nominal length-to-width-to-thickness ratio of 600:30:1. In the following micromechanics analysis the transversely isotropic ribbons therefore represent bundles rather than homogeneous glass material. In keeping with this bundle model, we shall use the subscript b on quantities related to the bundles. The subscript m is used to denote quantities for the matrix (which actually consists of resin filled with hard particles that are very small compared to the ribbons).

The Halpin-Tsai equations are approximations that relate effective elastic composite properties of a unidirectional composite to constituent values. Ac-

SYMBOL	TEMP(C)	STRESS(PA)	T PRIME(SEC)	CYCLE #	SPEC #
□	23.9	1.66200E+07	640	7	B16
○	23.9	3.33900E+07	640	7	B16
▲	23.9	5.01500E+07	640	7	B16
+	23.9	6.62500E+07	640	7	B16

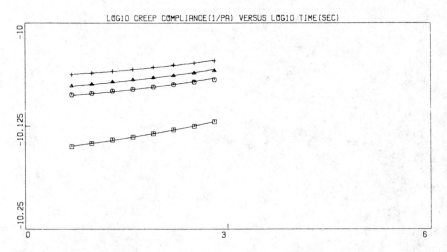

FIG. 3—*Creep compliance at 23°C for creep stress levels of 10, 20, 30, and 40 percent of the room-temperature ultimate strength.*

cording to this model, the longitudinal modulus E_1 and major Poisson's ratio ν_{12} are determined by the rule of mixtures:

$$E_1 = E_b v_b + E_m v_m \tag{3}$$

and

$$\nu_{12} = \nu_b v_b + \nu_m v_m \tag{4}$$

where E, ν, and v denote uniaxial modulus, Poisson's ratio, and volume fraction, respectively. The effective modulus E_2 for transverse, uniaxial loading is given by

$$\frac{E_2}{E_m} = \frac{1 + \zeta_E \eta_E v_b}{1 - \eta_E v_b} \tag{5}$$

where

$$\eta_E \equiv \frac{\lambda_E - 1}{\lambda_E + \zeta_E} \tag{6}$$

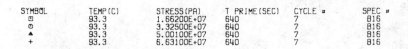

SYMBOL	TEMP(C)	STRESS(PA)	T PRIME(SEC)	CYCLE #	SPEC #
□	93.3	1.66200E+07	640	7	B16
⊙	93.3	3.32500E+07	640	7	B16
▲	93.3	5.00100E+07	640	7	B16
+	93.3	6.63100E+07	640	7	B16

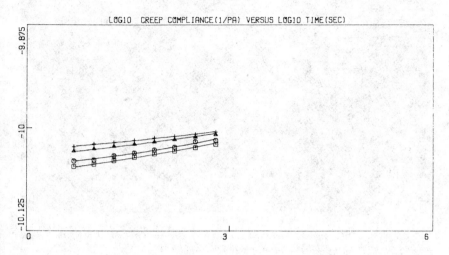

FIG. 4—*Creep compliance at 93.3°C for creep stress levels of 10, 20, 30, and 40 percent of the room-temperature ultimate strength.*

and

$$\lambda_E \equiv E_{2b}/E_m \tag{7}$$

in which E_{2b} is the transverse uniaxial modulus of the bundle. The principal shear modulus is

$$\frac{G_{12}}{G_m} = \frac{1 + \zeta_G \eta_G v_b}{1 - \eta_G v_b} \tag{8}$$

where

$$\eta_G \equiv \frac{\lambda_G - 1}{\lambda_G + \zeta_G} \tag{9}$$

and

$$\lambda_G \equiv G_{12b}/G_m \tag{10}$$

FIG. 5—*SMC-R50 composite after the resin has been partially oxidized, showing the orientation and structure of the ribbon-shaped bundles of glass fibers.*

For an isotropic matrix:

$$G_m = \frac{E_m}{2(1 + \nu_m)} \tag{11}$$

The parameters ζ_E and ζ_G depend on bundle geometry. Their values for several cases are established in Ref *12* by fitting G_{12} and E_2 to some numerical elasticity solutions for these effective moduli. The values of $\zeta_E = 2$ and $\zeta_G = 1$ seem to provide satisfactory results for many composites with continuous circular fibers. The ribbon-shaped bundles in the SMC composite are assumed to be continuous in length; a simple shear-lag analysis was made using the bundle and matrix properties deduced from the subsequent data analysis, and the correction to E_1 for finite length was estimated to be less than 5 percent. However, as the bundles are rectangular in cross section, the values of $\zeta_E = 2$ and $\zeta_G = 1$ are not necessarily the best ones. For *isotropic* ribbons, it is reported that [*12*]

$$\zeta_E = 2a/b \qquad \zeta_G = (a/b)^{1.73} \tag{12}$$

where a/b is the ratio of width to thickness. For the ribbons in Fig. 5, $a/b \simeq$ 30 and hence $\zeta_E = 60$ and $\zeta_G = 360$. Calculations of the SMC composite modulus E_c and Poisson's ratio ν_c using these values turn out to be virtually the same as those using the standard values for circular bundles; in fact, using realistic properties, these isotropic properties are found to be practically independent of the parameters over $0 \leq \zeta_E, \zeta_G < \infty$ because the transverse ribbon and composite properties are close.

Next, we use the laminate analogy [12-14] to relate the above unidirectional properties to E_c and ν_c for a random fiber composite; in this analogy, the composite properties are assumed equal to those for a quasiisotropic laminate. Following the notation and results in Ref 12, the composite E_c and ν_c will be expressed in terms of the invariants of the plane stress stiffness matrix, Q_{ij}. Specifically

$$E_c = 4U_5\left(1 - \frac{U_5}{U_1}\right) \tag{13}$$

and

$$\nu_c = 1 - 2\frac{U_5}{U_1} \tag{14}$$

where

$$U_1 \equiv \frac{1}{8}(3Q_{11} + 3Q_{22} + 2Q_{12} + 4Q_{66}) \tag{15}$$

and

$$U_5 \equiv \frac{1}{8}(Q_{11} + Q_{22} - 2Q_{12} + 4Q_{66}) \tag{16}$$

and where the Q_{ij} values are expressed in terms of the unidirectional composite moduli:

$$Q_{11} = NE_1 \qquad Q_{22} = NE_2$$

and $\hspace{10cm}$ (17)

$$Q_{12} = \nu_{12}NE_2 \qquad Q_{66} = G_{12}$$

with

$$N \equiv (1 - \nu_{12}\nu_{21})^{-1} \qquad \nu_{21} = E_2\nu_{12}/E_1 \tag{18}$$

Important limiting values of E_c and ν_c are those for a continuous, random fiber composite with matrix moduli E_m and G_m which are vanishingly small compared to the axial fiber or bundle modulus E_b. For this case it is obvious that $E_2 = G_{12} = \nu_{21} = 0$ (regardless of the particular micromechanics model used), and $E_1 = E_b V_b$. Equations 13 and 14 reduce to

$$E_c = \frac{E_1}{3} = \frac{1}{3} E_b V_b \qquad \nu_c = 1/3 \qquad (19)$$

where Eq 3 has been used for E_1.

The bundles in the SMC composite are themselves unidirectional composites consisting of isotropic, circular glass fibers in a presumably isotropic matrix. The mechanical properties of this matrix may be different from those for the matrix outside of the bundles; it is likely that the volume fraction of calcium carbonate particles is relatively small in the resin within the bundles and, in fact, may be essentially zero since the particles are not in the original fiber glass roving [15].

With this difference in mind, we shall designate the properties of the matrix within the bundles by primes in using the Halpin-Tsai equations to the predict the bundle properties. Thus, with subscripts f and m denoting (glass) fiber and matrix quantities, respectively,

$$E_b = E_f v_f + E'_m v'_m \qquad (20)$$

$$\nu_b = \nu_f v_f + \nu'_m v'_m \qquad (21)$$

$$\frac{E_{2b}}{E'_m} = \frac{1 + 2\eta'_E v_f}{1 - \eta'_E v_f} \qquad \eta'_E \equiv \frac{E_f/E'_m - 1}{E_f/E'_m + 2} \qquad (22)$$

$$\frac{G_{12b}}{G'_m} = \frac{1 + \eta'_G v_f}{1 - \eta'_G v_f} \qquad \eta'_G \equiv \frac{G_f/G'_m - 1}{G_f/G'_m + 1} \qquad (23)$$

where v_f and v'_m are the volume fractions of fiber and matrix, respectively, referred to the volume of a bundle. The values of $\zeta_E = 2$ and $\zeta_G = 1$ have been used as the glass fibers are circular. The fibers and matrix may be assumed isotropic, and therefore

$$G'_m = \frac{E'_m}{2(1 + \nu'_m)} \qquad G_f = \frac{E_f}{2(1 + \nu_f)} \qquad (24)$$

All glass fibers are assumed to be in bundles, and therefore

$$v_b = v_g/v_f \qquad (25)$$

where v_g is the volume fraction of glass referred to the entire composite. If we assume all calcium carbonate particles are outside of the bundles, the volume fraction of particles referring to only the matrix outside of the bundles is

$$v_{p0} = v_p v_f / (v_f - v_g) \tag{26}$$

where v_p is the volume fraction of the particles referred to the entire composite. Also

$$v_b + v_m = 1 \qquad v_f + v'_m = 1 \qquad v_r + v_p + v_g = 1 \tag{27}$$

where v_r is the volume fraction of resin in the composite; here, we consider the resin to include the polymer plus the small amounts of various other constituents such as catalyst, thickener, etc. [2].

In order to demonstrate certain features of the micromechanics model, we shall use the published volume fractions [2]:

$$v_g = 0.354 \qquad v_r = 0.540 \qquad v_p = 0.106 \tag{28}$$

and mechanical properties of the constituents:

$$\begin{array}{ll} \text{glass } [12]: & v_f = 0.22 \qquad E_f = 73 \text{ GPa} \\ \text{from our matrix tests:} & v_m = 0.35 \qquad E_m \simeq 10 \text{ GPa} \end{array} \tag{29}$$

The calcium carbonate particles are much stiffer than the resin, and thus micromechanics theory for a particulate composite predicts that E_m is essentially proportional to E'_m and that $v'_m \simeq v_m$ (see, for example, Ref 10). The volume fraction of fibers in the bundles, v_f, is needed in the calculations; from the experimentally estimated fiber count (200), ribbon cross-sectional dimensions (1.27 by 0.0423 mm) and fiber diameter (0.0127 mm) we find $v_f \simeq 1/2$. This value of v_f is only a rough estimate, as it is difficult to determine the bundle thickness; but it is of interest to note that Eqs 26 and 28 and the condition $v_{p0} < 1$ imply that $v_f > 0.4$. With $v_f = 1/2$, Eqs 25 to 28 yield

$$v_b = 0.708 \qquad v_{p0} = 0.363 \tag{30}$$

In contrast, if we change the ribbon thickness to 0.0282, then

$$v_f = 0.708 \qquad v_b = 0.500 \qquad v_{p0} = 0.212 \tag{31}$$

Whether we use Eq 30 or 31 has no effect on the *in situ* creep compliance, which is deduced in a later section from test data; therefore in this section we shall use only Eq 31 in an example.

Figure 6 shows the predictions of two different sets of micromechanics equations and two different microstructures as a function of matrix modulus. The left ordinate is the composite Young's modulus E_c less E_N, where

$$E_N \equiv \frac{1}{3} v_g E_f \tag{32}$$

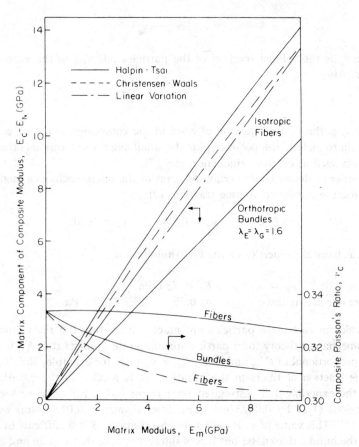

FIG. 6—*Micromechanics predictions of Young's modulus and Poisson's ratio for a quasi-isotropic laminate.*

is the modulus of a network of randomly oriented continuous glass fibers. The solid lines designated "fibers" were obtained from Eqs 13 and 14 after employing the Halpin-Tsai equations for a unidirectional composite with isotropic fibers (that is, the bundle model was not used). For this case, Christensen and Waals [16] derived E_c and ν_c, starting with more rigorous micromechanics equations for the unidirectional composite, and their results are also shown. Both theories predict that $E_c = E_N$ when $E_m = 0$; both are seen to be quite close for the range $0 \leq E_m \leq 10$ MPa. Notice that the dependence of ν_c on E_m from the Halpin-Tsai theory is not particularly good.

Of particular interest is the small amount of curvature, $d^2E_c/dE_m{}^2$, exhibited by both theories. It turns out that this curvature has a very pronounced effect on the *in situ* matrix modulus or compliance deduced from experimental data on E_c; indeed, using the theory without bundles and its extension to viscoelastic behavior, as described in the next section, the *in situ* matrix creep compliance is found to be not only different from data on unreinforced matrix but also qualitatively incorrect. On the other hand, good agreement is achieved from the bundle model, which turns out to predict that $d^2E_c/dE_m^2 \simeq 0$. (Although the curvature for isotropic fibers in Fig. 6 is not large, it has a major effect on the inferred time-dependence of matrix compliance.)

The solid lines in Fig. 6 for "bundles" were predicted using $\nu_b = 0.5$ (compare Eq 31) and $\lambda_E = \lambda_G = 1.6$; the value of 1.6 was selected to achieve agreement with the experimental data for the *in situ* and directly measured matrix material. Given λ_E and λ_G, it is not necessary to use explicitly Eqs 20 to 25 for predicting bundle properties. Indeed, these latter equations simply provide a means for estimating Young's modulus E'_m for the resin within the bundles; we found $E'_m \simeq 3$ GPa when $E_m = 10$ GPa. The former value is very reasonable for a glassy polymer without the calcium carbonate particles [17]. Also, it should be added that Eqs 20 to 25 result in values of λ_E and λ_G that are not quite equal, but whose average is 1.6 and only weakly dependent on E_m (assuming, as discussed previously, that E_m/E'_m is constant). The use of $\lambda_E = \lambda_G$ instead of the actual predicted values provides a prediction of ν_c that is in better agreement with experimental data. In view of the earlier comments concerning the Halpin-Tsai and Christensen-Waals predictions of ν_c for the fiber model (compare Fig. 6), it is believed the need to use $\lambda_E = \lambda_G$ is just a consequence of the approximate nature of the Halpin-Tsai equations; resolution of this point apparently must await development of a more accurate orthotropic ribbon model.

The most important findings from the ribbon model is that it predicts that λ_E and λ_G are essentially independent of E_m (at least for $E_m \ll E_f$) and, in turn, that $E_C - E_N$ is virtually proportional to E_m, as shown in Fig. 6. We have found this proportionality to exist for a great many cases involving various constituent volume fractions and properties and ribbon-width-to-thickness ratios, as long as λ_E and λ_G are *constant*. Thus

$$E_c = E_N + CE_m \qquad (33)$$

where C is independent of E_m. This equation, together with the subsequent viscoelastic analysis, provides the basis for obtaining the *in situ* creep compliance and establishing a method for predicting long-time viscoelastic response from short-time data. The numerical values of E_N and C will be found from analysis of the data, rather than from theory, considering the uncertainties in microstructural parameters, such as the particle locations and ribbon thickness. Also, the ribbons are not perfectly straight (compare Fig. 5), even though this is assumed in the theory.

Extension of the Theory to Linear Viscoelastic Behavior

When the composite is subjected to a constant load (creep test), the accurate quasi-elastic analysis method can be used to relate the *in situ* matrix compliance and Poisson's ratio to the corresponding quantities for the composite [10]. This method is valid even though the stress in the matrix is not constant. With reference to Eq 33, we obtain the relation between matrix creep compliance, $D_m = D_m(t)$, and composite creep compliance, $D_c = D_c(t)$, by simply replacing E_c with D_c^{-1} and E_m with D_m^{-1}. Thus

$$D_c = (E_N + CD_m^{-1})^{-1} \tag{34}$$

and thus

$$D_m = C(D_c^{-1} - E_N)^{-1} \tag{35}$$

As introduced in Eq 2 the recovery compliance D_r is equal to the strain following load removal divided by the creep stress. Through the superposition principle for linear materials, we obtain the relation between recovery and creep compliances for either the composite or the matrix:

$$D_r = D(t) - D(t - t') \tag{36}$$

where t' is the time of load removal (compare Fig. 1b).

The creep compliance of many materials can be accurately represented by the "generalized power law" [10]:

$$D(t) = D_0 + D_1 t^n \tag{37}$$

From Eqs 36 and 37:

$$D_r = D_1(t')^n [(1 + \lambda)^n - \lambda^n] \tag{38}$$

where

$$\lambda \equiv (t - t')/t' \tag{39}$$

Typically $D_1 t^n \ll D_0$ for short-time tests, and therefore accurate determination of n and D_1 from creep data is not possible; however, as described in Ref 6, creep and recovery data, together with Eqs 37 and 38, can be employed to obtain consistent, reliable values of D_0, D_1, and n.

A logarithmic form, $D = D_1 \ln(t/t_0)$, where t_0 is a constant, is often fit to creep data. We have found, however, that it can be used over only a relatively short time interval of about 15 min, whereas Eq 37 fits creep data over a much longer time; indeed this latter equation fits the matrix compliance over the entire range investigated. The logarithmic function cannot be expected to be very accurate, because it is only the first term in a power series expansion of a power law, namely:

$$(t/t_0)^n = 1 + n \ln(t/t_0) + \frac{1}{2}[n \ln(t/t_0)]^2 + \cdots \tag{40}$$

The Poisson's ratio of resin below the glass transition temperature is usually relatively constant [10], and therefore the Poisson's ratio of the composite for creep, $\nu_c = \nu_c(t)$, is predicted from Eq 14 after making the substitution $E_m \rightarrow D_m^{-1}$ and $\nu_m = 0.35$ (compare Eq 29). The quasi-elastic method can be used even if the creep Poisson's ratio is time-dependent by substituting $\nu_m(t)$ in place of the constant elastic value.

The linear viscoelastic two-dimensional, isothermal constitutive equations [referred to a rectangular Cartesian coordinate system for time-dependent stresses (σ_x, σ_y, τ_{xy}) and strains (ϵ_x, ϵ_y, γ_{xy})] can be written explicitly in terms of the composite creep properties D_c and ν_c [10]:

$$\epsilon_x = \int_{-\infty}^{t} D_c(t-\tau) \left[\frac{\partial \sigma_x}{\partial \tau} - \nu_c(t-\tau) \frac{\partial \sigma_y}{\partial \tau} \right] d\tau \tag{41a}$$

$$\epsilon_y = \int_{-\infty}^{t} D_c(t-\tau) \left[\frac{\partial \sigma_y}{\partial \tau} - \nu_c(t-\tau) \frac{\partial \sigma_x}{\partial \tau} \right] d\tau \tag{41b}$$

$$\gamma_{xy} = \int_{-\infty}^{t} J_c(t-\tau) \frac{\partial \tau_{xy}}{\partial \tau} d\tau \tag{41c}$$

where J_c is the shear creep compliance. The parameter J_c is related exactly to the time-dependent properties D_c and ν_c through an equation that is the same as for elastic media [10]:

$$J_c = 2(1 + \nu_c)D_c \tag{42}$$

In the remaining sections we will determine D_c and ν_c in terms of the *in situ* matrix compliance and parameters in the generalized power law for creep (Eq 37). Equations 41 and 42 then provide the generalization for any two-dimensional, time-dependent stress state in the linear viscoelastic range of behavior.

Experimental Procedure

Compression molded sheets of SMC-R50 furnished by General Motors Corporation were used to fabricate specimens with a 2.54 by 22.86 cm IITRI[4] tensile coupon geometry. Tabs of SMC-R50 were bonded to the grip section of the specimen with a structural adhesive. Final machining of each specimen was performed with a diamond saw to minimize edge damage.

Samples of the matrix system, which is a particulate-filled polyester, were fabricated at Texas A&M University by casting specimens in a precision mold. The specimens were cast in their final form in order to avoid mechanical damage to the specimens, which would occur if machining of the brittle matrix was necessary. After being cast in a precision silicon rubber mold, the samples were cured at 150°C for 15 min at a pressure of 0.7 MPa. By comparison, the nominal cure cycle for compression molded panels of SMC-R50 is 150°C for 2 to 3 min at approximately 7 MPa of die pressure. The long cure time of 15 min at 150°C used for the resin was necessary in order to allow the interior of the silicon mold to reach the cure temperature. The geometry was the same as that of the SMC samples except the end tabs were molded.

Three replicate creep and recovery experiments were performed on the SMC material at five equally spaced temperatures from 24 to 93°C and at four equally spaced stress levels from 10 to 40 percent of the room-temperature ultimate strength. The values used are shown in Table 1 for both composite (B16, B18, B2, B20) and matrix samples (R1).

Foil strain gages were used to measure strain during the experiments. The load was applied by a dead-weight lever-arm creep machine.

As noted previously, several cycles of loading and unloading were required before repeatable creep and recovery behavior was observed. The accelerated mechanical conditioning technique illustrated in Fig. 1a was used to minimize the time required to condition the sample while producing only a small amount of accumulated recovery strain. The time schedule for loading and unloading was a recovery period equal to twice the previous creep period. After the seventh cycle of this type of conditioning repeatable behavior was observed. Tables 1 and 2 summarize the test conditions and times.

Data Reduction

Experimental results will now be reported and the data used to evaluate parameters in Eqs 33 and 37. The response of one composite specimen will be discussed since its behavior is typical. The analysis is verified by comparing the *in situ* matrix compliance (the matrix compliance computed from the

[4]Illinois Institute of Technology Research Institute, Chicago, Ill.

TABLE 1—*Cumulative load history showing cycle number of creep and recovery for data used in characterization.*

Specimen	Creep Stress, MPa	Temperature, °C				
		24	41	57	77	93
B16	16	7	14	21	28	35
	33	63	...	...	...	42
	50	70	...	...	...	49
	66	77	...	...	...	56
B18	16	7	14	21	28	42
	33	70	...	...	...	49
	50	77	...	...	...	56
	66	86	...	...	...	63
B2	16	22	44	29	54	37
	33	5	...	...	...	76
	50	13	...	...	...	69
	66	16	...	...	...	83
B20	16	15	...	...	...	19
R1	4.6	10	12	13	16	18

TABLE 2—*Time (t', seconds) under load for cycle of creep and recovery used in characterization.*

Specimen	Creep Stress, MPa	Temperature, °C				
		24	41	57	77	93
B16	16	640	640	640	640	640
	33	640	...	...	...	640
	50	640	...	...	...	640
	66	640	...	...	...	640
B18	16	640	630	630	630	630
	33	630	...	...	...	630
	50	630	...	...	...	630
	66	630	...	...	...	630
B2	16	894	640	313	640	1283
	33	894	...	...	...	640
	50	892	...	...	...	640
	66	893	...	...	...	640
B20	16	507 600	...	...	...	606 402
R1	4.6	48 419	15 668	14 725	23 153	19 984

composite compliance and micromechanics) to the experimentally measured matrix compliance.

Creep and recovery compliances of the "resin" sample (which is actually the SMC material without the glass) at different temperatures are shown in Figs. 7 to 9. The stress level for these tests was chosen so that the strain was approximately equal to the strain in the composite during the lowest stress level experiment. Because the slope changes with time, neither a logarithmic law nor a simple power law (Eq 37 with $D_0 = 0$) can fit the data in Figs. 7 and 8, respectively. The solid lines are the best least squares fit of Eq 37 to the creep and recovery data; the parameters D_1 and n were first derived from the recovery data using Eq 38 and D_0 was derived from creep data using Eq 37.

A master curve (Fig. 10) was formed by vertical and horizontal shifting of the compliance versus log time curves at different temperatures. A generalized power law (the solid curve in Fig. 10) accurately predicts the compliance where, for a reference temperature of 366.3 K,

$$D_0 = 1.62 \times 10^{-10} \, (\text{Pa})^{-1}; D_1 = 1.82 \times 10^{-11}; n = 0.180 \qquad (43)$$

SYMBOL	TEMP (C)	STRESS (PA)	T PRIME (SEC)	CYCLE #	SPEC #
□	28.2	4.55200E+06	48419	10	R1
⊙	40.6	4.55200E+06	15668	12	R1
▲	57.2	4.55200E+06	46427	14	R1
+	76.6	4.55200E+06	23153	16	R1
×	96.6	4.55200E+06	19984	18	R1

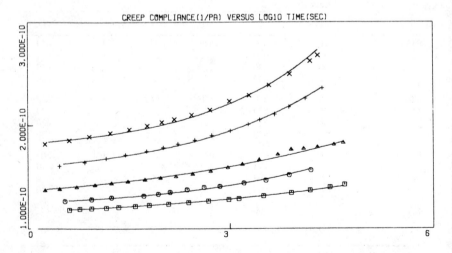

FIG. 7—*Creep compliance versus log time of resin alone as a function of temperature.*

SYMBOL	TEMP(C)	STRESS(PA)	T PRIME(SEC)	CYCLE #	SPEC #
□	28.2	4.55200E+06	48419	10	R1
⊙	40.6	4.55200E+06	15668	12	R1
▲	57.2	4.55200E+06	14725	13	R1
+	76.6	4.55200E+06	23153	16	R1
×	96.6	4.55200E+06	19984	18	R1

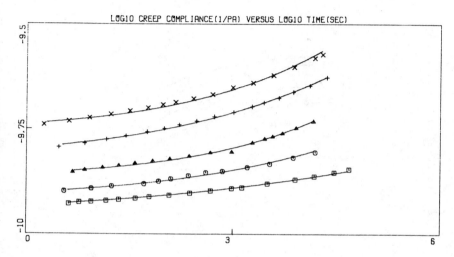

FIG. 8—*Log creep compliance versus log time of resin alone as a function of temperature.*

Only creep data were used to determine these constants. The recovery data are accurately predicted by means of Eq 38.

The fact that a master curve described by a generalized power law can be formed implies the matrix creep compliance as a function of time and temperature has the standard form [10]

$$D_m(t, T) = D_0(T) + D_1(t/a_T)^n \qquad (44)$$

where D_1 and n are the constants in Eq 43 and D_0 has the value in Eq 43 at $T = 366.3\,K$. The vertical shift required to form the master curve is equal to the change in D_0 (ΔD_0) between the test and reference temperatures; the horizontal shift equals log a_T, in which $a_T = a_T(T)$ is the so-called time-temperature shift factor; $a_T = 1$ at the reference temperature. Figure 11 gives both log a_T and ΔD_0. (Although we could have set $a_T \equiv 1$ and let D_1 be temperature dependent, standard practice is followed here; for transient temperature applications this option does not exist and one must determine if D_1 is a function of temperature [10].)

The composite creep compliance as a function of time and temperature was used to compute the *in situ* matrix compliance by means of Eq 35. The

SYMBOL	TEMP(C)	STRESS(PA)	T PRIME(SEC)	CYCLE #	SPEC #
□	28.2	4.55200E+06	48419	10	R1
⊙	40.6	4.55200E+06	15668	12	R1
▲	57.2	4.55200E+06	14725	13	R1
+	76.6	4.55200E+06	23153	16	R1
×	96.6	4.55200E+06	19984	18	R1

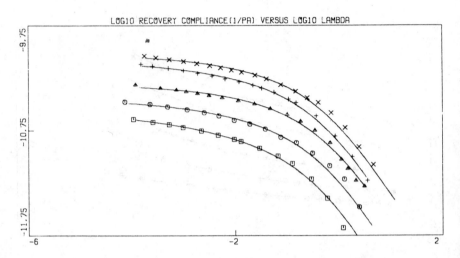

FIG. 9—*Log recovery compliance versus log time of resin alone as a function of temperature* $(\lambda = (t - t')/t')$.

constant E_N was estimated by equating the slope of the log-compliance log-time graphs of the *in situ* and actual matrix compliance at the highest temperature and lowest stress level. Note that the constant C does not influence this slope. Next, C was determined by equating the vertical shift between creep data at the highest and lowest temperatures on the log-compliance log-time graph of the *in situ* and actual matrix compliance. Table 3 shows the constants for all four samples. The micromechanics analysis predicts E_N, the modulus of a network of randomly oriented fibers, from Eqs 28, 29, and 32, to be 0.85 (10^{10}Pa), while the value determined from experimental data is 18 percent lower (Table 3). The experimental value of E_N is lower because the fibers within the bundles are not straight (Fig. 5), resulting in a reduced effective modulus of the network.

Figure 12 shows the resulting *in situ* matrix creep compliance versus log time. Just as with the matrix sample, it is found that a master curve can be formed by horizontal and vertical shifting of data; the shift parameters are shown in Fig. 11. The master curve constants are found to be the same as for the resin sample (Eq 43). However, note that all vertical and horizontal shifts are not quite the same. It is suggested that there may be a stress level de-

SYMBOL	TEMP(C)	STRESS(PA)	T PRIME(SEC)	CYCLE #	SPEC #
□	28.2	4.55200E+06	48419	10	R1
⊙	40.6	4.55200E+06	15668	12	R1
▲	57.2	4.55200E+06	46427	14	R1
+	76.6	4.55200E+06	23153	16	R1
×	96.6	4.55200E+06	19984	18	R1

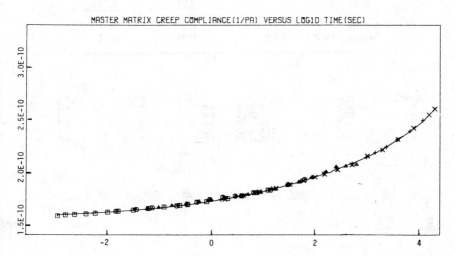

FIG. 10—*Master curve of resin creep compliance versus log time for a reference temperature of 366.3 K.*

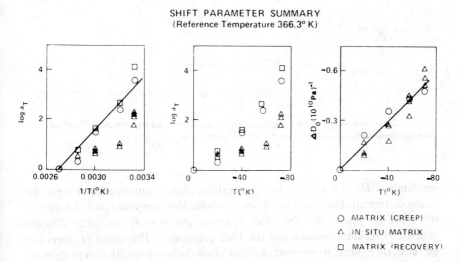

FIG. 11—*Shift parameters* a_T *and* ΔD_0 *for the matrix master curve. The value of* a_T *decreases and* D_0 *increases with increasing temperature.*

TABLE 3—In situ *matrix compliance parameters for four specimens.*

Sample	C_2	E_N (10^{10}Pa)
B-16	0.95	0.61
B-18	1.14	0.57
B- 2	0.93	0.76
B-20	0.93	0.78
Average	0.99	0.68
Average with B-18 deleted	0.94	0.72

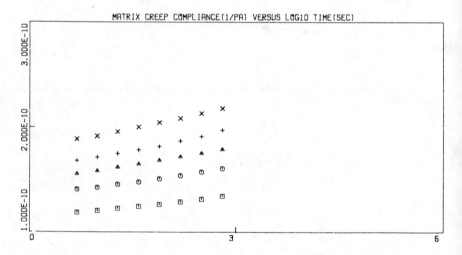

SYMBOL	TEMP(C)	STRESS(PA)	T PRIME(SEC)	CYCLE #	SPEC #
□	23.9	1.66200E+07	640	7	B16
⊙	40.6	1.66200E+07	640	7	B16
▲	57.2	1.64900E+07	640	7	B16
+	76.6	1.66200E+07	640	7	B16
×	93.3	1.66200E+07	640	7	B16

MATRIX CREEP COMPLIANCE(1/PA) VERSUS LOG10 TIME(SEC)

FIG. 12—In situ *matrix creep compliance versus log time at a creep stress of 10 percent of the room-temperature ultimate strength at varying temperatures.*

pendence of D_0 and a_T, just like that exhibited by a unidirectional composite in Ref 6. Indeed, the stress on the resin within the composite and the stress in the resin sample under the same external strain level are quite different because of residual stresses and the fiber constraint. The effect of stress level on the resin sample itself needs further study before the differences shown in Fig. 11 can be definitely related to stress effects.

The a_T shift factors for the matrix sample were determined from creep in

Fig. 7 and from recovery in Fig. 9. Since D_0 does not effect the recovery response, a_T is determined from a simple vertical shift in recovery; this may be seen by recognizing that Eq 38 still applies if we let $D_1 \rightarrow D_1/a_T^n$. These recovery values of a_T agree with the creep values of a_T determined by combined vertical and horizontal shifts.

Poisson's ratio for the composite was determined as a function of stress level from tension tests to failure at two rates, 44.5 and 445 N/s, and at three temperatures, 23.9, 57.2, and 93.3°C. A three-way analysis of variance of the data indicates negligible dependence of ν_c on temperature or loading rate. However, there is a dependence on stress level. The value of ν_c at the low stresses was 0.31 and it uniformly increased to a value of 0.27 at fracture. This agrees in the low stress region with the micromechanics prediction from Eq 29 and Fig. 6 that $\nu_c = 0.31$.

A test of the theory would be to predict the creep and recovery of the composite from the matrix creep and micromechanics theory. The creep response of the resin from Eq 37 and the constants from Eq 43 predict the creep response of the composite (Eq 34), using C and E_N from Table 3 and the appropriate shift factors for a given temperature from Fig. 11. Figure 13 shows

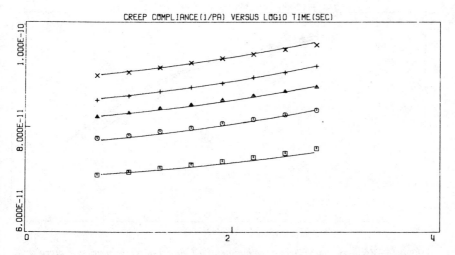

SYMBOL	TEMP(C)	STRESS(PA)	T PRIME(SEC)	CYCLE #	SPEC #
▫	23.9	1.66200E+07	640	7	B16
⊙	40.6	1.66200E+07	640	7	B16
▲	57.2	1.64900E+07	640	7	B16
+	76.6	1.66200E+07	640	7	B16
×	93.3	1.66200E+07	640	7	B16

FIG. 13—*Creep compliance of composite as a function of temperature at a creep stress equal to 10 percent of the room-temperature ultimate strength. The solid lines are the predicted compliances from the matrix creep and micromechanics theory.*

that composite creep can be accurately predicted. Predicting the recovery response of the composite using the analysis will provide an additional verification of the analysis, since recovery data were not used in determining the parameters of the model. Recovery is predicted in a similar manner except that Eq 38 is used in place of Eq 37. An accurate prediction of the composite recovery is shown in Fig. 14.

The parameters of the model may be determined by alternative approaches to those presented. For example, the constants C and E_N (Eq 34) are determined from creep experiments. An alternative means of determining the parameter E_N is from tensile failure data. The value of E_N corresponds to the modulus of a network of randomly oriented fibers that is also the modulus at failure during a tensile stress-strain experiment. The locus of failure points in a tensile stress-strain experiment for SMC-R50 at room temperature defines a straight line with a slope of 7.2 GPa [11]. This value of E_N agrees with E_N determined from creep (Table 3). Using this alternative approach to determine E_N leaves only C to be determined from creep of the composite.

SYMBOL	TEMP (C)	STRESS (PA)	T PRIME (SEC)	CYCLE #	SPEC #
□	23.9	1.66200E+07	640	7	B16
○	40.6	1.66200E+07	640	7	B16
▲	57.2	1.64900E+07	640	7	B16
+	76.6	1.66200E+07	640	7	B16
×	93.3	1.66200E+07	640	7	B16

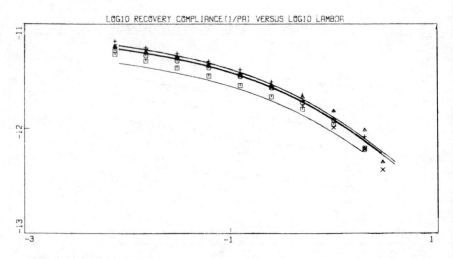

FIG. 14—*Recovery compliances of composite as a function of temperature at a creep stress equal to 10 percent of the room-temperature ultimate strength. The solid lines are the predicted recovery compliances from the matrix creep response and micromechanics theory.*

Conclusions

The time-dependent behavior of SMC-R50 in uniaxial tensile creep and recovery experiments is complex. This complex behavior is illustrated by data at different times and temperatures that cannot be related through a single reduced-time parameter and that indicate a decreasing creep rate at high stresses or temperatures or both. However, the composite response can be described, at least for low stress levels, by a master creep curve for viscoelastic response of the resin alone, micromechanics, and the laminate analogy. Although the analysis is quite involved and requires the introduction of ribbon-shaped bundles of fibers, the creep compliance of the composite D_c can be computed from the approximate theoretical relationship

$$D_c = (E_N + CD_m{}^{-1})^{-1}$$

where C and E_N are constants and D_m is the matrix creep compliance. The compliance data at higher stress levels indicate that nonlinear stress-dependent factors need to be included in the analysis.

Using short-term 4-h creep tests, the authors obtained a single master curve covering 10^6 s of reduced time for the creep compliance of the matrix. The predicted composite creep behavior as determined from the analysis and matrix master curve was experimentally checked through long-term (one-week) tests on the composite at ambient and elevated temperatures.

Acknowledgments

The authors gratefully acknowledge the financial support of this research by the Plastics Materials Characterization Department, Manufacturing Development Division, General Motors Corporation. The capable assistance of Dr. R. M. Alexander and Mr. C. Fredericksen in conducting the experimental program is greatly appreciated. Mr. M. McEndree and Mr. M. Kerstetter prepared the samples from the compression molded sheets of SMC-R50 supplied by General Motors Corporation.

References

[1] Heimbuch, R. A. and Sanders, B. A., "Mechanical Properties of Automotive Chopped Fiber Reinforced Plastics," in *Composite Materials in the Automotive Industry,* American Society of Mechanical Engineers, New York, 1978, p. 111.
[2] Denton, D. L., "Mechanical Properties Characterization of an SMC-R50 Composite," Paper 790671, in *Proceedings,* Annual Meeting, Society of Automotive Engineers, Detroit, 1979.
[3] Riegner, D. A. and Sanders, B. A., "A Characterization Study of Automotive Continuous and Random Glass Fiber Composites," Report MD79-023, Manufacturing Development Division, General Motors Corp., Warren, Mich., 1979.

[4] Struik, L. C. E., *Physical Aging in Amorphous Polymers and Other Materials,* Elsevier, New York, 1978.

[5] Cartner, J. S., Griffith, W. I., and Brinson, H. F., "The Viscoelastic Behavior of Composite Materials for Automotive Applications," in *Composite Materials in the Automotive Industry,* American Society of Mechanical Engineers, New York, 1978, p. 159.

[6] Lou, Y. C. and Schapery, R. A., *Journal of Composite Materials,* Vol. 5, 1971, p. 208.

[7] Schapery, R. A., *Journal of Polymer Engineering and Science,* Vol. 9, No. 4, 1969, p. 295.

[8] Beckwith, S. W., "Viscoelastic Characterization of a Nonlinear Glass/Epoxy Composite Including the Effects of Damage," Report MM 2895-74-8, Texas A&M University, College Station, Tex., Oct. 1974.

[9] Schapery, R. A., Beckwith, S. W., and Conrad, N., "Studies on the Viscoelastic Behavior of Fiber-Reinforced Plastic," Report MM 2702-73-3 (AFML-TR-73-179), Texas A&M University, College Station, Tex., 1973.

[10] Schapery, R. A., "Viscoelastic Behavior and Analysis of Composite Materials," in *Mechanics of Composite Materials,* Vol. 2, G. Sendeckyj, Ed., Academic Press, New York, 1974, p. 86.

[11] Alexander, R. M., Schapery, R. A., Jerina, K. L., and Sanders, B. A., this publication, pp. 208–224.

[12] Ashton, J. E., Halpin, J. C., and Petit, P. H., *Primer on Composite Materials: Analysis,* Technomic, Stamford, Conn., 1969.

[13] Halpin, J. C. and Pagano, N. J., *Journal of Composite Materials,* Vol. 3, 1969, p. 720.

[14] Halpin, J. C., Jerina, K. L., and Whitney, J. M., *Journal of Composite Materials,* Vol. 5, 1971, p. 36.

[15] Denton, D. L., Owens-Corning Fiberglas Corp., private communication.

[16] Christensen, R. M. and Waals, F. M., *Journal of Composite Materials,* Vol. 6, 1972, p. 518.

[17] Ferry, J. O., *Viscoelastic Properties of Polymers,* 2nd ed., Wiley, New York, 1970.

Summary

These papers on short fiber reinforced composite materials fall into three major areas: (*1*) materials characterization, (*2*) material modeling/property prediction, and (*3*) fracture behavior. The following sections summarize the papers from each of the three groups.

Materials Characterization

The mechanical properties of three short fiber materials are investigated by *Walrath et al*. Their work represents a start on generating a basic material property data base for two appearance grade automotive short fiber composite (SFC) materials (SMC-R25 and SMC-R30) and for one structural grade material (SMC-R65). Mechanical properties tested in their study included tensile strength and modulus, compressive strength, flexural strength and modulus, shear strength, and impact strength. Results for all three materials indicated that tensile and compressive properties decrease with increasing temperatures. For the tension specimens, the properties also decreased with increasing moisture absorption. However, a combination of absorbed moisture and elevated temperature did not cause additional significant degradation. Results from the short beam shear test verified that this test method is not a good one for generating design numbers for shear, since both bending and shear failures are obtained. Flexural properties from both the three-point and four-point test modes were very similar. Of the two impact methods used (tensile and Charpy), tensile impact data offered the more reliable results. Charpy results were dependent on sample orientation and were useful for relative ranking of the three materials but not for quantitative design values. The authors were able to obtain useful conclusions and trends of behavior of the three materials systems under the test conditions in their program.

Gibson et al concentrate primarily on the dynamic properties of short fiber composites. They present base line data for small amplitude vibration at various frequencies under room temperature and humidity conditions. The materials tested included two suppliers' appearance grade systems (SMC-R25 from PPG and OCF), neat resin, and three structural grade systems (SMC-R65, SMC-C20/R30, and XMC-3). The latter two materials are composed of both continuous and chopped fibers. Complex moduli were measured using a forced flexural vibration technique. The flexural mode was selected because it represents the most common mode of structural vibration these materials would see in automotive structural application. The complex moduli of all the materials were found to be essentially independent of frequency and amplitude within their test ranges, thus permitting the use of complex moduli notation for analysis. In general, materials tested having the greatest storage moduli had the lowest loss factors, while those with the lowest storage moduli had the highest loss factors. This was attributed to whether the materials are matrix controlled or fiber controlled. The vibration damping properties of these short fiber composite materials were greater than those for aluminum.

Collister and Gruskiewicz also address dynamic mechanical characterization of SFC materials. They investigate the dynamic mechanical properties of a commerically available SMC system by looking at the properties of the constituent systems during various stages of the molding cycle. The results of their study provide insight into the effect the various additives have on the molding compound at different stages of molding.

Duke addresses material characterization from a quality control vantage point. He studies two SMC-R65 systems with different glass length and type (1 and 2 in. E-glass and 1 and 2 in. S-glass). He evaluates several nondestructive evaluation (NDE) methods, including ultrasonic C-scan, X-ray radiography, vibrothermography, and acoustic emission (AE) monitoring in order to relate the NDE findings to the required mechanical property performance. If appropriate defect-property correlation can be made, NDE techniques offer the potential of identifying and screening among the fabricated-in defects to determine which will be detrimental to the materials performance in service. The property used for correlation studies was static tensile stength. The author found that some of the NDE techniques investigated could complement each other, while others could not be readily correlated with material performance. Ultrasonic C-scan showed the capability of identifying planar defects; vibrothermography tended to correlate with C-scan results. AE monitoring indicated that damage may be occurring very early during loading the SMC-R65 materials. X-ray radiography did not produce conclusive results with respect to correlation to material performance, but it did offer the potential of being useful for examining specimen fiber distribution. The author indicates that a more extensive study needs to be done to determine the full potential of correlating NDE techniques parameters to the mechanical response of the SFC materials.

Owen reviews static and fatigue property of chopped strand mat (CSM)/ polyester resins (PR) composites. The paper represents a compilation and review of the various data collected by Owen and his associates on CSM/PR systems. The fatigue and static strength of CSM/PR composites were studied under axial and bi-axial loading. Various aspects of failure processes, cumulative damage, crack propagation, and fracture mechanics were investigated under these loading conditions. The data generated were reviewed for their use in a coherent design methodology for CSM/PR material systems. Test results showed significant scatter in properties; this leads to very conservative engineering designs. Owen also noticed that there is no clear-cut endurance limit for CSM/PR composites as for many engineering metals and that the strength properties show a size effect dependence. Although further work needs to be done, it appears that fracture toughness and macroscopic crack growth may be a more reilable approach to strength prediction.

Jerina et al look at the tensile creep and recovery behavior of SMC-R50. Their study involves both experimental and theoretical investigations. They measured the strain at different stress levels and temperatures and compared this to the predictions of micromechanics theory. This type of work is essential for providing the understanding and confidence required in the material's long-term strength and dimensional stability. Test results show a decreasing creep rate at high stresses or temperatures or both. The authors were able to predict the creep and recovery of the composite from the matrix creep experimental data and micromechanics calculation of the effective properties of the composites.

Tung departs from generating basic mechanical properties alone by investigating the effects processing conditions have on the properties of a short fiber composite. The effect of varying such processing parameters as cure time, mold temperature, and mold pressure is investigated on mechanical properties (tensile, flexural, impact properties) and on a thermal property (heat deflection temperature). Results showed that varying processing conditions had a more pronounced effect on the thermal property than on the mechanical property investigated.

Material Modeling/Property Prediction

Three papers address material modeling/property predictions. *Caulfield* developed a model to predict the modulus and the coefficient of thermal expansion (CTE) of the chopped glass fiber/epoxy system over cryogenic temperature ranges (300 to 78 K). Both Young's modulus and CTE vary with temperature in this temperature range. A set of constitutive equations is established for strain response over the indicated temperature range. Test results show close correlations between the proposed time-dependent constitutive equations predictions and experimental results.

Chang uses an analytical model to evaluate the effect on stiffness of SMC composites by varying fiber content, modulus, and density. Results of this analytical variation show that the stiffness of chopped fiber SMC composites can be increased by increasing fiber content and fiber modulus or decreasing fiber density. This study provides valuable information on material design tradeoffs to achieve higher stiffness materials than with conventional glass SMC composites.

An analytical approach for predicting properties for structural SMC systems is presented by *Sridharan*. He uses laminate theory models to predict elastic properties of various chopped/continuous SMC composite combinations. The chopped and continuous portions of the SMC systems are considered as individual plies with either solely chopped or solely continuous layers. Experimental results are in full agreement with analytical predictions for stiffness predictions but less so for strength predictions. The analytical models presented here and in the prior two papers may prove a significant advancement in short fiber composites technology, because they offer the potential of confidently screening for new material development. Better models to more accurately predict strength properties would help minimize much of the extensive testing required on the continually new materials development coming from suppliers.

Fracture Behavior

Three papers address various aspects of fracture behavior of SFC materials systems. *Mandell et al* look at the mode of crack propagation in injection molded short glass and carbon fiber reinforced thermoplastics. The matrix materials used included nylon 66, polycarbonate, polysulfone, and others covering a range of very brittle to ductile systems with 30 to 40 percent fibers. The authors found that the main crack in each material appeared to follow a fiber avoidance mode, that is, growing around longer fibers and agglomerations of locally aligned fibers. Although additional work needs to be done to address material behavior with cracks grown in other than normal to the dominant fiber direction, the results of this work offer useful direction in material development to improve material performance.

Alexander et al investigated fracture behavior of a structural short glass fiber system, SMC-R50, at varying loading rates and under different environmental condition. Several different crack lengths were used with samples of two different widths. Varying the load rate showed that loading rate had no significant effect on ultimate tensile strength. Though the stress-strain curve did show a nonlinearity, it was possible to characterize the fracture stress behavior by using linear elastic fracture mechanics (LEFM) theory and a constant fracture toughness. In order to use LEFM theory in design with SMC-R50, it was necessary to replace ϵ by ϵ^p in predicting overall sample strain. The authors acknowledge that more data need to be generated to fully validate their findings.

Another discussion of fracture of short fiber composites was presented by *Wang and Yu*. They discussed an analytical method for studying fracture initiation based on the Weibull statistical strength theory. In their analytical model they incorporate random orientation of the fiber mats and structural heterogeneity of the composite microstructure. Results were analyzed based on the thermomechanical loading condition, and suggest that the computer-aided simulation scheme used may be suitable for studying statistical fracture initiation.

Final Remarks

The papers in this volume cover a broad area of research and development activities needed in SFC materials. Much of the work presented produces results that show trends rather than a large statistical data base. Continuing R&D activities on these material systems should provide the data and understanding of behavior needed to build the confidence of designers in more widespread use of short fiber composites.

B. A. Sanders

General Motors Manufacturing Development,
GM Technical Center, Warren, Michigan;
editor

Index

A

Acoustic emission monitoring, 106–112

Aerospace applications, 64, 133, 225

Agglomerations, fiber, 6, 9, 47

Analysis (*see* Testing, Thermal mechanical analysis technique, Ultrasonic C-scanning, Vibrothermography, X-radiography)

ASTM Test for

Apparent Interlaminar Shear Strength of Parallel Fiber Composites by Short Beam Method (D 2344), 123

Compressive Properties of Unidirectional or Crossply Fiber-Resin Composites (D 3410), 121

Deflection Temperature of Plastics under Flexural Load (D 648), 53

Flexural Properties of Plastics and Electrical Insulating Materials (D 790), 52, 124, 135

Impact Resistance of Plastics and Electrical Insulating Materials (D 256), 52

Notched Bar Impact Testing of Metallic Materials (E 23), 126

Plane-Strain Fracture Toughness of Metallic Materials (E 399), 5, 209

Tensile Properties of Fiber-Resin Composites (D 3039), 170

Tensile Properties of Plastic (D 638), 4, 52, 98, 116

B

Asymmetric four-point bending (AFPB) shear test, 124

Automotive structures, composites in, 33, 50, 113, 151, 225, 226

Vibration characteristics of, 133–149

Bending, cause of material failure, 123, 135

Bending stiffness, 39

Boltzmann superposition principle, 89, 91

Bonding, resin-fiber, 59

Bond strength, 4, 9

Bulk molding compounds (BMC), 151, 183, 184

C

Calcium carbonate ($CaCO_3$), 34, 35, 50, 115, 169, 174, 185, 186, 194, 197–200, 202–205, 226, 234, 235, 237

Carbon fibers, 8, 9, 34, 35, 42, 44, 46, 47

Carbon-reinforced materials, 5

Failure of, 29

Fracture surfaces in, 24, 25

Polycarbonate, 9

Charpy impact tests (*see* Testing, Charpy impact)

Chopped fiber reinforced sheet molding compound composites (*see* Sheet molding compounds)

Chopped-mat fiber composites, randomly oriented, fractures in, 151–165
Cracks (see also Flaws, Fracture)
Advance of, local mechanisms of, 8–21
Coalescence mode, 20
Growth, 20, 65, 68
Fiber avoidance mode of, 6–8
In short-fiber composites, 152
Of CSM/PR laminates, 80–82
Length, 223
Variation with fracture stress, 217
Variation with tensile strength, 214, 216
Path, 8, 14, 15, 20
In glass-reinforced PAI, 23
Propagation, modes of, 3–32
Resistance, 9
Of glass fiber composites, 209
Of short fiber composites, 30
Tip zone, 8–11, 20
Creep
Compliance, 229–231, 237, 238, 242–244, 246
Properties, 184, 210, 223, 226, 227, 240–243
Rupture, 223
Stress, 229–231, 238, 247, 248
Tests, 90, 221, 223, 238, 249
Uniaxial tensile, 249
Variation with stress, 227
Variation with temperature, 227, 244, 247
Creep-recovery periods, 228
Cryogenic fluid transport and storage vessels, 152, 153

D

Damping, 134, 135, 137, 147, 148, 185
Measurement of, 139
Variation with amplitude, 142
Variation with frequency, 140
Thermoelastic, 137, 141
Debonding, 9, 20, 21, 29, 31, 66, 68, 73–76, 209
Deformation, 145, 221
Degradation of material, 130
Due to prolonged curing, 61, 62
Rate of, 28
Structural, of tensile properties, 55
Density, composite, 35, 38, 40–42, 44, 46–48
Diffusion behavior of materials, Fickian, 120, 121

E

Elastic behavior, linear, 227–237
Elastic modulus, 36, 37, 44, 184
Of HSMC, 177, 179, 180
Of SMC, 114, 115, 117, 118, 125, 135, 170
Elastic properties
Continuous/chopped glass fiber hybrid SMC, 167–181
Polymers, 195
SMC-R50, 229
Elastic shear modulus, polyester, 185, 187, 191, 206, 207
Energy
Absorption, 114, 128, 130
Loss, 141
Entanglement coupling, 188, 191, 192
Epoxy, 87, 88
Chopped strand and continuous fiber glass/epoxy, 28
Crack growth in, 209
Matrix composites, 121, 161
Extensometer data, 87, 170, 210, 211, 214, 215

And strain gage, 113, 116, 121, 214, 215

F

Failure, material, 9, 20, 102, 104, 109, 130
Early, 78
In fiberglass, 107
Mechanisms of, 65-68, 72
Patterns, 100, 108, 110, 111
Probability, 161, 162
Specimen, 98, 99, 103, 104
States of, 75
Strength, 155
Stress, 163, 209, 210, 220, 222, 223
Tensile, 123, 130
Fatigue, 5, 69, 71-75, 76, 78
Cracks, 20, 21
Cycling, 20
Effects, 21, 28, 29
Loading, 66, 68
Properties, 5
Resistance, 3, 8, 29, 31
Tests, 66
Fiber
Avoidance mode, 6-8, 21
Bending, 30, 31
Density, effects of, 40-43, 46
Failure, 9, 20
Length, 4, 28, 30, 31, 218, 223
Modulus, 34, 35, 46
Effect of, 40-42
Orientation of, 9, 115, 118, 124, 126, 130, 155, 156, 160, 162, 164
Pullout, 8, 9, 20, 21, 209
System, 33-48
Type, effects of on material stiffness, 42-48
Volume fraction, 38, 39
Weight fraction, 34, 38, 39, 46
Filler, 34-36 (*see also* Calcium carbonate)

Density, 38
Inert, 50
Of polyesters, 196-200
Properties, 42
Flaws
Effect on material fracture, 156, 157, 163, 165
Growth of, 103-105, 108-111, 208, 226
Pre-existing, 208, 209, 216
Size of, 223
Flexural properties, 56, 57, 60, 61
Fracture
Initiation, 151-159
Simulation procedure, 159-161
Mechanics, 65, 76
Linear elastic, 209, 214-220
Of chopped-mat fiber composites, 151-165
Probability of, 157, 158
Random fiber composites, 208-223
Surface, 20
Toughness, 3, 4, 8, 21, 28, 30, 68, 80, 82, 208, 209, 214, 216, 218, 223
Frequency, effects on materials, 135, 140, 144-147, 194-196

G

Glass
Content, effects on HSMC properties, 170
E-glass, 51, 52, 98, 108, 109, 134-149, 174
Fiber, 35, 40, 42, 44, 46, 47, 186
Bundles, 232, 234
Chopped, 64-83
Fiber reinforced composites, 4, 5, 20, 64-82, 198, 199, 200, 204, 205
Crack growth, mode of, 28
Crack resistance of, 31
Failure of, 28, 32

Fracture surfaces in, 26, 27
Nondestructive characteristics of, 97–111
SMC, 39, 40, 44, 46, 47
Fiber-reinforced plastics (GFRP), 65, 73, 86, 88, 94, 95, 226
S-2 glass R65, 110, 111, 174
Graphite-reinforced materials, 31, 32, 59, 174, 226

H

Halpin-Tsai equations, 227, 229, 234–237
Heat deflection temperature (HDT), 53, 54, 61, 62
Hybrid sheet molding compound (HSMC)
Elastic modulus, 179, 180
Elastic properties of, 172
Failure of, 174, 177
Strength of, 180
Tensile strength of, 172–176

I

Impact (*see also* Testing, Charpy impact)
Direction, 128
Strength, 126, 128
Velocity, 126, 129

K

Kaiser effect, 107, 111, 112
Kevlar, 34, 35, 40, 42, 44, 46, 47

L

Laminates
Analogy, 169, 170–173, 233, 249
Failure of, 174, 177
Glass chopped strand mat/polyester resin
Crack growth in, 80–82
Failure mechanisms in, 65–68, 72
Fatigue in, 69, 71, 72, 74, 75
Fracture toughness, 80, 82
Glass content, 70
Strengths of, 76–79
Stress concentration, 75, 76
Stress rupture, 69, 73, 74
Variabilities of properties, 68, 69
Micromechanical predictions of, 236
Properties of, 172
Stiffness of, 169
Strength of, 169
Stress-strain response of, 169
Unidirectional-ply, 5
Lamination theory, 227

M

Magnesium oxide, 185, 186, 189, 192–194
Mechanical testing, dynamic (*see* Testing)
Micromechanics theory, 221, 223, 227–249
Moisture absorption, 114, 117–119, 130
Effects on elastic modulus, 117, 118
Molecular weight, effects of in polyester, 188, 191–193, 195, 207

N

Nondestructive evaluation (NDE) techniques, 98–107
Comparison of, 108–110, 209
Nylon 66 (N66), 4, 5, 25, 27
Carbon-reinforced, 32
Crack growth in, 7, 17, 18, 20

P

Peroxide, testing of, 185, 186
Plastics (*see also* Thermoplastics)
Fiber reinforced (FRP), 65, 75, 76, 133, 142
Glass reinforced (GRP), 65, 73

Poly(amide-imide) (PAI), amorphous, 5, 6, 9, 20
  Crack path in, 19, 24
  Damage development in, 22, 23, 26
Polycarbonate (PC), 4, 5, 20
  Carbon-reinforced, 25
  Cracks in, 14, 15
    Growth of, 7
  Glass-reinforced, crack tip in, 18, 27
Polyester composites
  Alkyd, 187, 188, 190, 191, 199
  Cross-linked polyester resin, 187, 197, 202, 203
  Dynamic mechanical characterization of, 183–207
  Filled polyester compound, 187, 194, 196–198
  Filled reinforced polyester compound, 187, 196, 198–200
  Molding applications of, 193
  Resin, 34, 35, 50, 51 (see also Laminates, Sheet molding compounds)
    Glass reinforced, 210
    Thickened, 187, 189, 191–193
  Rheological characterization of, 193
  SMC, E-glass reinforced, 113
Polymers
  Fiber-reinforced, applications of, 225, 226
  Thermoset, 53, 55
    Breakdown of, 54, 55
Polystyrene blends, 192, 193
  Frequency for, 195, 196
Polysulfone (PSUL), 4, 5, 16

R

Recovery properties of SMC-R50, 227, 240–244, 247–249
  Function of temperature, 248
Resin, 34–36 (see also Polyester composites)

Cracking of, 66, 73–79
Density, 38
Matrix, 137, 145, 147, 148, 152, 161
Polymeric-based neat, 50, 53
Properties, 42
Unsaturated polyester, 185
Volume fraction, 38–41, 46, 235
Weight fraction, 34, 38

S

Semicrystalline polyphenylene sulfide (PPS), 5
  Carbon-reinforced, 20
    Cracking in, 12, 13, 24
  Fatigue resistance in, 29
  Glass-reinforced, 10, 11, 20, 26
  Matrix, 8, 9
Sheet molding compounds (SMC) (see also Hybrid sheet molding compound), 5, 33–48, 98, 183, 184
  Effects of fiber systems on stiffness properties of, 33–48
  Filled, 34–37, 40, 42, 43, 46–48
  Mechanical behavior of, 113–131
  Mechanical properties, 52, 53
    Flexural properties, 60–62
    Impact property, 61
    Processing variables, 54–61
    Studies of, 51
    Tensile properties, 54–59
  Molding conditions, 51, 52
  Processing parameters, 53–62
  Processing techniques, 51
    Variables of, 51–63
  Thermal properties, 53, 54
    Effect of processing variables on, 61, 62
    Studies of, 51
  Types of
    SMC-30, 40, 47
    SMC-40, 41
    SMC-45, 41

SMC-50, 34, 40
SMC-65, 40, 47
SMC-R25, 51–53, 59, 61–63, 113–131
SMC-R30, 113–131
SMC-R50, 209–223, 228–249
SMC-R65, 98, 113–131, 170–172, 176
Unfilled, 34, 38, 40–46
Uses of
  Automotive industry, 151, 226
  Transportation industry, 167
Stiffness, 33, 61, 135, 142, 143, 145, 147, 148
  Continuous fiber material, 167
  Dynamic, 134, 135
  Effects of fiber systems on, 33–48
  HSMC, 179–181
  Longitudinal versus transverse, 137
  Matrix, 160
  SMC-R50, 214, 227
Strain
  Amplitude, maximum, 139, 141–143
  Capability, in PAI, 6
  Dependence on stress-time-temperature, 86–95
  Energy, 219
  Evaluation of, 161
  Maximum principal, 77
  Measurement of, 87–89, 240
  Thermal, 86
  To failure, 66, 67
Strength (*see also* Ultimate compression strength, Ultimate tensile strength)
  Compressive, 114, 121–124
  Fatigue, 65–82
  Material, 5
  Moisture absorption, 118–120
  Shear, 114, 124, 125
  SMC, 167–181
  Stiffness, 97
  Tensile, 116, 118, 120, 123

Tensile impact, 114
Ultimate, of SMC-R50, 212, 229–231, 240, 246–248
Weight, 50
Stress (*see also* Failure, Thermomechanical stress) 71, 72, 74–77, 90–93
  At fracture, 215
  Concentration, 75, 76, 80
  Controlling of, 89
  Failure, 82, 99, 109
  Hoop, 76–79
  Intensity, 209, 210, 214, 216
    Critical, 218, 219
  Levels, 69, 245–249
  Local, 21
  Material, 30, 31
    Complexity of, 131
  Rate tests, 90, 91
  Rupture, 69–75
  Shear, 20, 111
  State, 164, 239
  Stress-strain behavior, 91–96
    Of HSMC, 180
    Of SMC-R50, 209, 211, 213–215, 220, 221
  Tensile, 248
  Stress-temperature relationship, 91, 92, 94, 95
  Styrene, 184, 186, 201

T

Temperature (*see also* Heat deflection temperature)
  Cryogenic, 85, 95
  Effect on compressive strength, 122, 123
  Effect on creep compliance, 227, 230, 231
  Effect on elastic modulus, 117
  Effect on material strain, 87–95
  Effect on material tensile strength, 57–59, 116–120, 123
  Glass transition, 239

Tensile failure, 69, 248
Tensile impact, 125–128, 130, 131
Tensile modulus of SMC, 117, 119, 121
Of HSMC, 171, 177, 178
Tensile strength
Of CSM/PR laminates, 76, 78
Of HSMC, 172–176
Of PAI, 6
Of SMC, 51–59, 221
Crack length, variation with, 214, 216
Effect of cure time and pressure on, 53–55
Tension (see Testing)
Testing
Charpy impact, 114, 125, 126, 128–131
Environmental, 141
Impact, 125–130
Mechanical
Dynamic, 184–200
Of SMC materials, 113
Pressurized leakage, 161
Shear, 123, 124
Solid, of polyesters, 187–189
Static compression, 121–123
Static tension, 115–120
Tension, 29
Quasi-static, 110
Static, 115–120
To failure, 247
Uniaxial, 6, 9, 156
Thermal mechanical analysis technique, 53
Thermal properties, materials, 141, 145
Of SMC, 53, 54, 61, 62
Thermal shock, 153, 154
Thermoelastic loss factor, 145
Thermomechanical loading, biaxial, 151–165
Thermomechanical stress, 153, 156–159, 162, 164

Thermoplastics
Crack propagation in, 3–32
Engineering, 5
Injection molded fiber reinforced, 3–32
Thickening agent, chemical, 185, 193, 200, 207

U

Ultimate compression strength (UCS), 68, 70
Ultimate tensile strength (UTS), 4, 21, 28, 30–32, 65, 66, 68–70, 82, 99, 170, 179, 210
Ultrasonic C-scanning, 100–104, 106–108, 110

V

Vibration characteristics of composites, 133–149
Vibrothermography, 106, 107, 108, 110
Vinyl ester, Dow 790, 169, 174
Viscoelastic behavior, linear, 238–249
Viscoelastic matrix behavior, 145, 147
Viscoelastic properties
Of polymers, 191
Of SMC-R50, 223, 225–249
Viscoelastic theory, 188
Viscosity
Effects on material properties, 184, 185, 189, 191, 193, 194, 196, 200, 207
Non-Newtonian, 194

X

X-ray radiography, 104–106, 108, 111

Z

Zener thermoelastic theory, 139
Zinc stearate, 185, 186